***BETWEEN
ALCHEMY
AND
TECHNOLOGY***

BETWEEN ALCHEMY AND TECHNOLOGY

The Chemical Laboratory

JUDITH A. WALMSLEY

FRANK WALMSLEY

Department of Chemistry
The University of Toledo
Toledo, Ohio

PRENTICE-HALL, INC. Englewood Cliffs, New Jersey

Library of Congress Cataloging in Publication Data

Walmsley, Judith A
 Between alchemy and technology.

Includes bibliographical references.
 1. Chemistry—Laboratory manuals.   I. Walmsley,
Frank, joint author.   II. Title.
QD45.W23   542   74-31447
 ISBN   0-13-075945-7 pbk.

© 1975 by PRENTICE-HALL, INC.
Englewood Cliffs, New Jersey

All rights reserved. No part of this book
may be reproduced in any form or by any means
without permission in writing from the publisher.

10 9 8 7 6 5 4 3 2 1

Printed in the United States of America

PRENTICE-HALL INTERNATIONAL, INC., *London*
PRENTICE-HALL OF AUSTRALIA, PTY. LTD., *Sydney*
PRENTICE-HALL OF CANADA, LTD., *Toronto*
PRENTICE-HALL OF INDIA PRIVATE LIMITED, *New Delhi*
PRENTICE-HALL OF JAPAN, INC., *Tokyo*

Contents

	Preface	xi
	A Message to You from the Authors	xiii

Part I TECHNIQUES

A	**Safety**	3
B	**Data**	6
	1. Recording Data	6
	2. Graphical Treatment of Data	6
	3. Significant Figures	9
	4. Errors, Precision, and Accuracy	10
C	**Balances**	16
	1. Trip Balance	17
	2. MultiBeam Balance	17
	3. Top Loading Balance	17
	4. Analytical Balance	19
	5. Buoyancy Correction	21
D	**Weighing Samples**	23
	1. Methods of Weighing	23
	2. Quantitive Transfer of Solids and Solutions	23
	3. Drying Samples	25
E	**Cleaning of Glassware**	27
F	**Liquid Measure**	29
	1. Beakers and Erlenmeyer Flasks	29
	2. Graduated Cylinders	29
	3. Volumetric Glassware: Flasks, Pipets, Burets	30
	4. Standard Solutions and Primary Standards	33
	5. Use of Volumetric Ware in Experiments	34

	G	**Filtration**	**36**
		1. Decantation	36
		2. Gravity Filtration	36
		3. Suction Filtration	37
	H	**Heating**	**40**
		1. Methods of Heating	40
		2. Glass Bending	41
		3. Heating Liquids and Solutions	44
	I	**Instrumentation**	**45**
		1. Magnetic Stirrer	45
		2. Centrifuge	45
		3. pH Meter	46
		4. Spectrophotometer	47
	J	**Mélange**	**51**
		1. Reading a Vernier	51
		2. Use of a Wash Bottle	51
		3. Inserting Glass Tubing into a Rubber Stopper	52
		4. International Systems of Units	52
		5. Computers	54

Part II EXPERIMENTS

(weighing and measuring)

1	**Measuring Volume by Counting Drops**	**57**
2	**Calibration of Volumetric Glassware**	**61**

(separation and identification)

3	**Melting Points**	**65**
4	**Formula of a Hydrate**	**71**
5	**The Chemistry of Nitric Acid**	**79**
6	**Paper Chromatography**	**85**
7	**Thin-Layer Chromatography**	**93**
8	**Ion-Exchange Chromatography**	**99**
9	**Chemical Mystery Theater**	**111**

(synthesis)

10	**Synthesis of a Coordination Compound**	**115**
11	**Synthesis of Tin(IV) Iodide**	**123**
12	**Polymers**	**129**
13	**Synthesis of an Amino Acid**	**137**

(thermochemistry)

14	**Thermochemical Cycle**	**145**
15	**Determination of a Simple Phase Diagram**	**151**

16	Heat of Fusion	159
17	Molecular Weight Determination by Freezing Point Depression	167

(quantitative determination)

18	An Acid-Base Titration	173
19	Titration Using a pH Meter	179
20	Volumetric Determination of Metals	183
21	Vitamin C Determination	189
22	A Gravimetric Analysis	199
23	Nitrogen Content of an Amino Acid	205
24	Spectrophotometric Analysis of Copper	211

(equilibrium and kinetics)

25	The Chromate-Dichromate Equilibrium	217
26	Dissociation of an Iron(III) Complex	223
27	Hydrolysis of a Tertiary Halide	231

APPENDIXES

I	Answers to Exercises	241
II	Computer Programs	257
III	Tables	267
	Constants	267
	Conversion Factors	267
	Salt Solubilities	268
	Periodic Table of the Elements	269

Preface

This manual has been written to accompany an introductory chemistry course for students whose professional goals are in the sciences, medicine, pharmacy, or engineering. We have chosen the experiments to acquaint the students with some of the newer techniques available today as well as some of the more classical techniques that are still important in laboratory chemistry. Even though the experiments are not the "traditional" ones for introductory chemistry, they are written for the average student using a non-calculus textbook. The experiments have been tested by several hundred students in our own classes at the University of Toledo, and their suggestions were used to improve the experiments.

Our aim has been to have the experiments teach important techniques and the principles upon which those techniques are based. For example, thin-layer chromatography is a widely used technique and it would be very tempting to be oriented toward the health sciences and separate amino acids. However, we have chosen the separation of some straight chain monocarboxylic acids because this provides a simple enough system to make correlations between the observed separations and the molecular structures of the compounds being separated. The apparatus has been kept as simple as possible and alternatives are often suggested. In Experiment 23, the student is able to do, in a simplified and somewhat less precise form, the classical Van Slyke method of determining the nitrogen content of amino acids. Although some experiments resemble those of a quantitative analysis course, they are not designed for the same degree of rigor. The spectrophotometric copper analysis (Experiment 24) does not involve preparation of a standardization curve but uses Beer's law to arrive at a direct relationship between absorbance and concentration. The gravimetric analysis (Experiment 22) is a system which gives a curdy, easily handled precipitate more suitable to beginners than silver chloride which tends to be fine and colloidal. The use of Gooch or filter crucibles is easier than the charring and burning of filter paper as done in a barium sulfate precipitation.

The amount of material included in the introductions to the experiments varies with the particular topic. For topics which are likely to be found in the student's text we have kept the introduction short. For topics not usually found in general chemistry texts, we have tried to be more complete. Since the order of topics differs from text to text, there may occasionally be mention of material unfamiliar to the student. In some cases this may be skipped over for the time being; in other cases this will present an opportunity to introduce the student to a new topic. Whichever choice is made in such instances, the learning of the main principles and the techniques of the experiment should be unaffected.

To assist students to do as much as they can on their own, we have written the laboratory techniques, the use of instruments, and other general information separately from the experiments. This should make it easier for a student to find out how to do something in general which can then be applied in various situations throughout the experiments. Within each experiment, the appropriate page or section number is given to help the student in finding the necessary information. For the students (particularly the average students) exercises with answers have been included with the experiments. The exercises ask questions to help the student think about how he/she is going to do the experiment and once it has been completed, how to do the necessary calculations. Just the presence of such exercises is usually insufficient because the student who can answer them easily has little need of them and the student who cannot answer them finds them of no value whatsoever. Thus we have included detailed answers to provide assistance to the student who needs that assistance and is willing to use this form of instruction. The true test of whether or not the student learned the material comes in the writing of the laboratory report and in the writing of examinations.

The laboratory is a good place to teach some descriptive chemistry and we have included experiments which do that in a (hopefully) painless but effective way. In particular, we have designed several experiments which involve reactions of metal ions in aqueous solution with an opportunity to use these reactions in the identification of unknowns in each experiment (Experiments 4, 5, 6, 8) and in a final nonstructured set of unknowns (Experiment 9).

The experiments are flexible. It is not necessary to do all experiments or to do them in any particular order. Some experiments are related such as the synthesis (Experiment 10) and analysis (Experiments 20, 22) of coordination compounds and the synthesis (Experiment 13) and analysis (Experiment 23) of glycine. However, if these syntheses are not carried out by the student, the analyses may still be performed using commercial unknowns and commercial glycine. In many instances, we have suggested alternative apparatus. For example, in Experiment 20, directions are given for adjusting the pH using either a pH meter or pH paper. For students without a powder funnel (Experiment 12) we suggest a rolled up piece of paper. Experiments that use a buret have directions for use of either a 25 or 50 ml buret.

We have tried to keep in mind the fact that the materials available to your students may vary, and for this reason we have chosen simple apparatus such as beakers and Erlenmeyers as much as possible. The choice of chemical systems was made partly on the basis of cost; the chemicals used are either inexpensive or are used in very small amounts. The laboratory report sheets have been kept short and each item numbered. This should help the students organize their data, direct them toward the important aspects of the use of that data, and make grading easier. Our experience has been with three-hour laboratory periods and the experiments reflect that experience. However, some experiments require more than one period for completion of all parts. Instruments used in these experiments are pH meters and spectrophotometers (Bausch and Lomb Spectronic 20 or equivalent). Volumetric glassware and analytical balances are required. Items such as these are expensive but we believe such items are really necessary for adequate training of today's students.

In an endeavor such as the writing of this manual there are many people who contribute. If it were not for them, this manual would be less than what it is and so we owe our special thanks to our chemistry department colleagues and to:

Kathy and Susan—for being interested in and enthusiastic about their parents' textbook;

Dr. S. Y. Tyree, Jr.—for fostering in us the spirit of academic excellence and an appreciation of the rewards of teaching;

Dr. Jack G. Kay and Dr. C. E. Griffin—for their cooperation and encouragement;

Sandra Flick and Joan Kent—for overcoming difficulties only typists know about.

JUDITH A. WALMSLEY
FRANK WALMSLEY

A Message to You from the Authors

We enjoy chemistry; yet we realize that everyone does not share our enthusiasm. As a matter of fact, there are areas of chemistry that even we find dull or difficult. We do not ignore these areas, however, because they are necessary for us to function as chemists in our chosen area. You are studying chemistry either because you enjoy it or because it will be useful in your chosen profession. Either way, this book will be of most value to you if you know how to use it and do use it. This message, then, is concerned largely with providing a summary of what useful features this book contains and how they may be best utilized.

In order to accomplish anything in a laboratory, it is necessary to know what equipment to use and to master the use of that equipment. Part I contains practical information on techniques and equipment. For example, you will not find electrical circuit diagrams of instruments but instead what knobs to look for and what they do. You will also find a section on safety (a *very important* topic) and even one on how to wash dishes. You don't learn a technique or how to use an instrument by reading about it, however; you have to do it, probably many times, before you become proficient. The experiments in Part II provide opportunities to learn by doing. The separation of the techniques from the experiments will make it easier for you to locate this information.

Although the emphasis in a laboratory course should be on laboratory techniques, the concepts utilized in the experiments are important and are briefly discussed in the introduction to each experiment. Use your text to study the concepts and models, and use the experiment introductions to provide bridges between the text and the actual performance of the experiment. You will notice that exercises are provided for each experiment. Some of these will help you in the actual performance of the experiment. Others will help you in doing the calculations that are necessary to the utilization of the primary data. Since your instructor will not be in your home or your dorm room when you sit down to write the report, we have provided detailed answers to the exercises. Once you can do the exercises, you should be able to do the report calculations.

Most of the experiments contain references that fall into two categories. The general references are books and articles that might be helpful to you if you want another source to study in addition to this manual and your text. We have attempted to limit these references to those which are not only appropriate and well written but also which you are likely to find in your school library. In addition, there are specific references that are the original reports of experimental work. It is unlikely that you will need to look up any of these references but they are listed if the need or curiosity does arise.

As you begin doing these experiments, you might ask whether or not they are really relevant to today's problems in health, ecology, consumerism, etc. If you feel that to be relevant you must go out and determine the sulfur dioxide content of air or the lead content of polar ice, then this manual is not relevant. We believe that our relevancy is long range, however, because the *specific* problems change but the experimental methods for solving problems remain basically the same. Thus, as you go on in your studies and as you go on to your chosen profession, you may encounter some of these same techniques again.

We hope your laboratory experience is an adventure in the sense that an adventure is an exciting experience. Webster's dictionary also defines an adventure as an undertaking of uncertain outcome and undoubtedly the results of some of your experiments are at this time not predictable. We also hope that you will treat each adventure as an endeavor. We choose the word *endeavor* because it means an exertion of intellectual strength toward the attainment of a goal. In other words, your laboratory experience is only going to be worthwhile if you work at it. The opportunity is here.

*BETWEEN
ALCHEMY
AND
TECHNOLOGY*

Part I
TECHNIQUES

A
Safety

When we (the authors) were undergraduates in chemistry, we were sometimes jokingly asked the question, "Well, did you blow up the lab today?". Fortunately, the answer was always "No." The chemical laboratory is a place where accidents can and do occasionally happen, a majority of them relatively minor. If a few simple safety rules are followed, however, the laboratory is probably a far safer place to be than riding in a car on our nation's streets or highways.

The basic safety rules for the chemical laboratory are

1. Always wear eye protection.
2. Avoid contact with chemicals (on the skin, by inhalation, or by oral ingestion).
3. Keep flammable chemicals away from open flames.
4. Do not smoke, eat, or drink in the laboratory.
5. Wash hands thoroughly at the end of the laboratory period.
6. Think before you act.

Some elaboration will probably help you understand the underlying reasons for the rules above.

Eye protection. The purpose of eye protection is to prevent splashing chemicals, shattered glass particles, or other flying objects from getting in your eyes. You will be required to wear some type of eye protection in the laboratory at all times, whether or not you are working on an experiment at a particular moment because others around you may be performing an experiment. Safety glasses (with plastic or tempered glass lenses) and goggles are the most frequently used types of eye protection in the general chemistry laboratory. Goggles and glasses with side shields have the advantage of protecting the eyes from objects approaching sideways but the disadvantage of limiting peripheral vision. Prescription eye glasses are generally considered adequate protection. All those manufactured after January, 1972 have heat-tempered lenses. Persons wearing contact lenses must use the same eye protection as those persons who do not normally wear glasses.

Contact with chemicals. Nearly all the chemicals that you will use in the laboratory are toxic to some extent and under certain conditions. Even sodium chloride is toxic if a person consumes an excessively large quantity of it. Pure water is one of the few chemicals that is completely nontoxic under all conditions. A large number of solid compounds are poisonous only if they are swallowed. Some solids and liquids have a high enough vapor pressure at room temperature (a reasonable number of molecules can exist in the gaseous state) so that the fumes or vapors of the chemical could be dangerous if it is a toxic chemical. Some compounds, largely organic liquids, can be rapidly absorbed through the skin. Therefore, as a general precautionary measure, contact with chemicals should be avoided or kept to a minimum.

Flammable chemicals. Many chemicals, especially organic compounds, will burn when ignited; however, the organic liquids pose the greatest fire hazard. Some of the liquids are very volatile and it is possible for the vapor to ignite even though the liquid is some distance from a flame. Spilled flammable liquids also ignite very easily. Therefore, keep all flammable chemicals away from open flames, sparks, or other sources of fire. Never heat a flammable liquid with a burner.

Smoking, eating, and drinking. No smoking is allowed in the laboratory because it poses a fire hazard. One should not eat and drink in the laboratory because it is possible that the food could be accidentally contaminated with chemicals.

Wash hands at the end of the period. Even though you believe your hands are free of chemicals, you should wash them just before leaving the laboratory for the day.

Think first, then act. A great many accidents could be prevented if everyone were to think before taking action. Probably all of us have said at some time in our lives, "If only I had stopped to think, I would have known better than to do such a thing!" In the laboratory it is imperative that you think about what you are going to do *before* you act. You must continue to think about what you are doing *while* you are doing it.

Chemical Spills. (a) In the eyes. If any chemicals get into the eyes, wash immediately with a gentle stream of water, holding the eyelids apart. Continue washing for at least 15 minutes and then call a physician. Some laboratories are equipped with "eye fountains" for this purpose but any source of water will do. Do not use boric acid ointments or other substances in the eyes in these cases. *If you wear contact lenses it is very important that you remove them before washing out the eyes.* Chemicals can become trapped behind them.
(b) On the skin. If you spill chemicals on your skin, whether they are harmful or not, wash them off with water immediately and thoroughly. Further treatment will depend on the type of chemical and if it causes burns. Ask your instructor's advice. Even concentrated acids and bases may not seriously burn you if they can be washed off quickly and thoroughly. A safety shower should be located in or near the laboratory for washing off major spills.
(c) On the clothing. If chemicals are spilled on your clothing, it may be necessary to remove the article of clothing, especially if the chemical was an acid or base or if a large quantity of any chemical was spilled. Use the safety shower in or near the laboratory if necessary. In all cases, report to your instructor immediately.
(d) On the bench top, balances, etc. You are expected to clean up any small chemical spills that you cause and you are provided with a sponge for this purpose. Clean up all spills *immediately* for safety reasons and as a matter of courtesy to your fellow students and instructor. This includes chemicals spilled on the balances, balance tables, reagent shelves, bench tops, and floors. If a large quantity of a chemical is spilled, consult your instructor concerning the best method to clean it up. Large quantities of concentrated acids and bases need to be neutralized first.

Fires. Many times small fires can be easily extinguished by smothering them with a nonflammable material. For example, a fire contained in a beaker or flask can be extinguished by placing an asbestos pad or watch glass on top of it. A fire extinguisher should be located in the laboratory for putting out larger fires.

Burns. (a) Thermal Burns. Most thermal burns can be prevented by "thinking before acting" and by using the proper equipment, i.e., tongs or gloves when handling hot apparatus. Cold water is considered an effective first-aid measure for small burns; the treatment should be continued until the pain subsides.

(b) Chemical Burns. As stated earlier in "Chemical spills," all chemicals that come in contact with the skin or clothing should be immediately and thoroughly washed off with water. Further treatment will depend on the particular chemical involved.

SPECIFIC REFERENCES

1. N. V. STEERE, ed., "CRC Handbook of Laboratory Safety", 2nd ed., The Chemical Rubber Co., Cleveland, Ohio, 1971.
2. Manufacturing Chemists Association, "Guide for Safety in the Chemical Laboratory", 2nd ed., Van Nostrand Reinhold Co., New York, N.Y., 1972.

B
Data

1. RECORDING DATA

Scientists have found it important to record their data and observations in a permanent form as quickly as possible. The record must be complete, accurate, and legible in order that results of an experiment can be correctly interpreted and in order that the experiment can be repeated if necessary. Many scientists record their data, observations and calculations in a bound notebook.

In order to assist you in keeping an accurate record of your laboratory work, this manual has spaces provided on the report sheets for you to record your data. Do not trust your memory; record the information immediately. Since scraps of paper and the like are liable to be lost, enter all your data and observations directly on the report sheet. Be careful to record what you have done or what you have seen and not what you think you should have done or what you think you should have seen. In addition to having an accurate record of your data, it is important that any results calculated from the data be easily checked. Therefore you should indicate how you did the calculations.

Care should be taken to separate the actual observation from the interpretation of the observation. For example, in 1900, the French chemist Victor Grignard in describing the results of an experiment initially wrote in his notebook "... en même temps qu'il se dépose MgI_2 en trés faible quantité" (at the same time MgI_2 precipitates in a small amount). He crossed out "dépose MgI_2" and replaced it with "produit un dépôt brun" (produces a brown deposit). His *observation* was a brown precipitate; his *interpretation* was that the brown solid was magnesium iodide but he had no direct evidence to confirm the interpretation. (Incidentally, his interpretation was not correct and to this day chemists are uncertain about the identity of that brown solid.)

2. GRAPHICAL REPRESENTATION OF DATA

Graphs have many advantages to favor their use in representing data. They may reveal maxima, minima, inflection points, or other significant features that might be overlooked in tabular or formula representation. Such features may also be overlooked by a computer treatment of data if the data are not exactly what was expected in advance. Chemists seem to have a penchant for plotting their data in a manner that yields a straight line. This may require some manipulation such as plotting y versus $1/x$ instead of y versus x or plotting log y versus x

instead of y versus x. This is not merely a whim of chemists but results from the fact that the slope of a line is a rate of change of the function plotted as ordinate with respect to the function plotted as abscissa; such rates are frequently physical properties of the system under study.

Some of the more important steps to be followed in preparing a satisfactory graph and in determining slopes of straight lines are given here. These suggestions are not inflexible and, in case of doubt, common sense should prevail.

1. *Choosing the graph paper.* There are numerous types of rectangular coordinate paper available ranging from 1 millimeter per division to 1/4 inch per division. The paper to be used depends on the range of values to be included on each axis and the accuracy with which these are to be plotted. Either 10 divisions per inch or 10 divisions per centimeter are satisfactory for most purposes. If one or both of the functions are logarithmic, semilog (y axis is a logarithmic scale) and log-log paper (both axes are logarithmic scales) are available.

2. *Choosing the coordinate scales.* (a) The independent variable should be plotted as the abscissa; the dependent variable, as the ordinate. The abscissa does not need to be the longer side of the graph; it could be the shorter side depending on the range covered by the variables. (b) Mark out scales such that as much of the graph paper is used as possible. That is, if your range of one variable is from 0 to 10 and there are 20 major divisions on the graph paper, use a scale of 2 major divisions for each unit of your variable rather than 2 major divisions for 2 units, which would result in using only half the graph paper. The uncertainties of measurement should not correspond to more than 1 or 2 of the smallest divisions.

3. *Labels.* The axes should be labeled at each or alternate main coordinate lines with the numerical value they represent. The name of the quantity represented and the units in which it is measured is to be given along each axis. A caption should also be included to tell what the graph is intended to show. The caption is usually placed in an open region on the graph paper unless it is to be reproduced for printing.

4. *Plotting the data.* Each point should be placed on the graph with a dot surrounded by a suitable symbol such as a circle. If more than one curve is to be plotted on the same graph, a different symbol should be used for each set of data. Symbols commonly used are a circle, square, triangle, and inverted triangle. When all points are plotted for a set of data, a smooth curve should be drawn through the points using a straightedge if the curve is a straight line or a French curve or spline if the curve is otherwise. It is not necessary that the curve pass through all the points but it should pass as closely as reasonably possible to all the points except those that appear to be grossly in error. In any event, the result should be a smooth curve. An example of a straight line graph is shown in Fig. 1; the data are given in Table 1.

5. *Determining the slope of a straight line.* The slope of a straight line is the amount of change in the dependent variable per unit change in the independent variable, $\Delta y/\Delta x$. If we choose two points on the line and designate the points as 1 and 2, then the slope of the line is $(y_2 - y_1)/(x_2 - x_1)$. In Fig. 1 two such points have been chosen and their values indicated with dotted lines. The slope of that line is $\Delta(\log P)/\Delta(1/T)$ or $(2.200 - 2.920)/(0.00323 - 0.00308) = (-0.720)/(0.00015) = -4.8 \times 10^3$. The negative sign indicates that the line slopes to the left, away from the vertical. The exact slope depends on how the line is drawn through the points. There is a mathematical procedure that gives a less arbitrary fit, called the method of least squares, which when applied in this case gives a slope of -4.71×10^3.

The data plotted in Fig. 1 are a study of the equilibrium

$$C_6H_5NH_2 \, (l) + SO_2 \, (g) \rightleftharpoons C_6H_5NH_2 \cdot SO_2 \, (s)$$

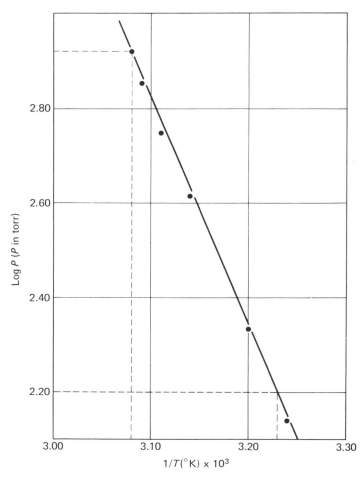

Figure 1. Aniline — sulfur dioxide equilibrium.

where the measured pressure is the equilibrium constant K_p and the slope of the line is ΔH for the reaction between aniline and sulfur dioxide. Converting the slope given above to kilocalories, one obtains $\Delta H = -21.3$ kcal. This particular example was chosen to illustrate how a graphical procedure gives useful information.

TABLE 1

Data for Aniline-Sulfur Dioxide Equilibrium

$t(°C)$	$T^{-1}\ (°K^{-1})$	P (torr)	log P
51.5	0.00308	832	2.920
50.2	0.00309	719	2.857
48.1	0.00311	561	2.749
45.4	0.00314	415	2.618
39.2	0.00320	216	2.335
35.0	0.00324	138	2.140

6. *Determining the end point of a titration curve.* A typical titration curve is shown in Fig. 2. It is composed of three regions, each of which approximates a straight line; these are labeled on the graph and extended out to show the points of intersection. Lines *A* and *B* intersect at point *D*, which corresponds to 4.10 ml. Lines *B* and *C* intersect at point *E*, which corresponds to 4.25 ml. The end point is the volume of titrant added halfway between points *D* and *E*, which is (4.10 + 4.25)/2 = 4.18 ml.

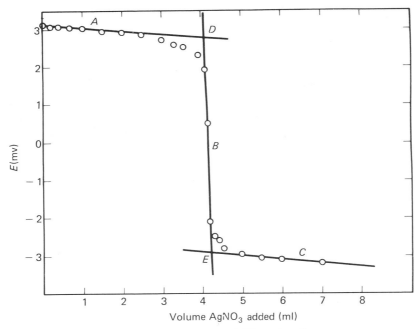

Figure 2. A Potentiometric Titration Curve.

3. SIGNIFICANT FIGURES

In reporting numerical information the data should contain the number of significant figures justified by the experimental method. When we are taking data directly from an instrument or a piece of apparatus, there is usually little doubt as to how the number should be reported if we are using the equipment correctly. Such data are called *primary data* because they are the first obtained; other data may then be derived from the primary data. For example, buret readings of 0.05 ml and 20.77 ml are primary data; the volume of solution, 20.77 ml − 0.05 ml = 20.72 ml, is secondary data.

TABLE 2
Examples of Counting Significant Figures

	Volume (ml)	*Volume (ml)*	No. of Significant Figures
1.	1.23	1.23	3
2.	3.560	3.560	4
3.	0.398	3.98×10^{-1}	3
4.	0.0213	2.13×10^{-2}	3
5.	987	9.87×10^{2}	3
6a.	250	2.50×10^{2}	3
6b.		2.5×10^{2}	2

Table 2 lists some numbers (volumes) as they might be found written down. In all cases except the last the number of significant figures can be easily determined, especially if the number is written in standard exponential notation. The second number has four significant figures (3, 5, 6, and 0); the zero is counted because it really means *zero*. In the fourth number there are only three significant figures (2, 1, and 3). The zero to the right of the decimal point is not a significant figure because it is being used only to designate the location of the decimal point; indeed, when written in standard exponential form, that zero disappears (into the exponent). If people were to write exactly what they meant, the sixth number would present no problem; it would have three significant figures (2, 5, and 0). However, numbers that have zeros immediately before the decimal point are often intended as approximate numbers; the zeros are only placing the decimal point. For example, one talks about a 250 ml beaker. This does not mean that such a beaker holds between 249 and 251 ml. It only means that such a beaker holds approximately 250 ml. Therefore when you see such numbers, proceed with caution.

Another case where the number as written may belie the true number of significant figures is the case of exact numbers. In these numbers, there is no uncertainty and therefore the number of significant figures is very large, larger than an experimental number. For example, we may look into a laboratory drawer and see four beakers and we would write "4 beakers." There are *exactly* four beakers; i.e., 4.0000000000... beakers. Another example is a conversion factor with a given system of measure. There are *exactly* 12 inches in 1 foot and *exactly* 100 centimeters in 1 meter.

When data are used in calculations, we should check our answers to see that the number of significant figures reported is consistent with the data used to obtain that answer. The following guides should be followed.

1. For multiplication and division, the answer should contain the same number of significant figures as the number used that contains the smallest number of significant figures. Example: (29.1) × (3.6982) = 108, which is 107.61762 rounded off to three significant figures.
2. For addition and subtraction, the guide in words becomes rather long and involved. It is best seen by taking an example: 22.5 + 0.22. Since the second decimal place is not known for 22.7, then the second decimal place cannot be known in the answer either.
3. For taking logarithms, the answer should contain *in the mantissa* the same number of significant figures as there are in the number itself.[1]
4. For taking antilogarithms, the answer should contain the same number of significant figures as there are in the mantissa.[1]

SPECIFIC REFERENCE

1. D. E. Jones, *J. Chem. Educ.*, 49, 753 (1972).

GENERAL REFERENCES

W. L. Masterton and E. J. Slowinski, "Mathematical Preparation for General Chemistry", W. B. Saunders Co., Philadelphia, Pa., 1970, Chap. 4.

C. H. Sorum, "General Chemistry Problems", 4th ed., Prentice-Hall, Inc., Englewood Cliffs, N. J., 1969 pp. 25-28.

4. ERRORS, PRECISION, AND ACCURACY

It's a fact of life that we all make mistakes. In a chemical laboratory these may range from *gross* errors like spilling half the solution to smaller errors caused by inexperience. These errors are usually obvious and can be corrected. Other such errors, called *systematic* or *determinate* errors, can also be corrected but are not always obvious. For example, a buret used at one time by one of the authors delivered to the 40 ml mark only 39.90 ml. If this buret had not been calibrated, all measurements using the buret in the vicinity of the 40 ml mark would have been in error by about 0.25%.

Even if all goes well, if everything is calibrated, and if we've done the experiment many times so that we're experienced and make no gross errors, subsequent measurements of the same quantity will not agree exactly. For example, three determinations of the concentration of a cerium (IV) sulfate solution gave results of 0.3930 M, 0.3916 M, and 0.4012 M. Such errors are called *random* or *indeterminate* errors.

If the only errors are random errors, we can judge the precision of our experiments and make an estimate of the accuracy as well. In addition, this will help us to identify a measurement where determinate errors may have crept in without our being aware of it. In the discussion following we shall define the terms *accuracy* and *precision* and see what use we can make of these in evaluating our results.

Accuracy and precision. These terms are often used interchangeably but should not be. Accuracy is a measure of the correctness of a measurement, that is, how closely a measurement agrees with the true value. Admittedly it is at times difficult to decide on a "true" value but if a true value is known, we can assess our accuracy by making a comparison of our result with that true value. In making such an assessment, it is customary to calculate the error and the relative error [expressed either as percent (%) or as parts per thousand (ppt)].

$$\text{Error} = \text{measured value} - \text{true value}$$

$$\text{Percent error} = \frac{\text{error}}{\text{true value}} \times 100$$

$$\text{Error in parts per thousand} = \frac{\text{error}}{\text{true value}} \times 1000$$

Precision is a measure of the reproducibility of a measurement and as such does not give any indication of error in the sense defined above. Precision cannot be directly related to accuracy without a clearly stated and understood assumption, namely that all errors are random errors.

If a measurement within a set of n measurements is denoted by X_i, it is possible to calculate for that set of measurements several useful quantities. The average (or arithmetic mean) $\overline{X}$ is the sum of all the measurements divided by n.*

$$\overline{X} = \frac{\sum_{i=1}^{n} X_i}{n}$$

The deviation of a measurement from the mean d_i is the difference between that measurement and that mean.

$$d_i = X_i - \overline{X}$$

The average deviation from the mean $\overline{d}$ (sometimes referred to as the precision because strictly speaking the average deviation is zero) is the sum of the deviations taken without regard to sign divided by n.

$$\overline{d} = \frac{\sum_{i=1}^{n} d_i}{n}$$

Since a relative average duration makes it easier to evaluate the magnitude of an absolute average deviation, it is common to calculate such a quantity.

$$\text{Relative } \overline{d} = \frac{\overline{d}}{\overline{X}}$$

This may be expressed as percent by multiplying by 100 or as parts per thousand by multiplying by 1000.

Standard deviation. The variance σ^2 is calculated by summing the squares of the

---

*A line over an algebraic designation means that it is an average value. Verbally it is read as *bar*; i.e., $\overline{X}$ is X bar.

deviations and dividing that sum by $n-1$. More commonly used than variance is the standard deviation σ.

$$\sigma = \left(\frac{\sum_{i=1}^{n} d_i^2}{n-1}\right)^{1/2}$$

If all errors are random, the standard deviation can be used to estimate the reliability of the data. This is stated in terms of the probability of the true value lying within a range surrounding the average value. The greater the range, the greater the probability of the true value being somewhere within that range. For example, we can be 100% sure that the true value lies somewhere between $-\infty$ and $+\infty$; if the range is less than that, our certainty becomes less. Table 3 summarizes some probabilities which were calculated assuming that the random errors are of a type which will yield a Gaussian distribution of values.

TABLE 3
Probability of a Result Falling within a Range of $\pm w\sigma$ around the Mean X

Range	Probability (%)
$X \pm 0.674\sigma$	50.0
$X \pm 1.00\sigma$	68.3
$X \pm 1.50\sigma$	86.6
$X \pm 1.65\sigma$	90.0
$X \pm 2.00\sigma$	95.5
$X \pm 2.58\sigma$	99.0

EXAMPLE 1. For the following set of numbers, calculate (a) average ($\overline{X}$), (b) average deviation ($\overline{d}$), and (c) standard deviation (σ).

3.864, 3.884, 3.799, 3.849, 3.849

Number	Deviation from Average	$\left(\begin{array}{c}\text{Deviation}\\ \text{from Average}\end{array}\right)^2$
3.864	0.015	0.00023
3.884	0.035	0.0012
3.799	0.050	0.0025
3.849	0.000	0.0000
3.849	0.000	0.0000
$\overline{X} = 5\overline{)19.245} = 3.849$	$\overline{d} = 5\overline{)0.100} = 0.020$	

$$\sigma = \left(\frac{0.0039}{4}\right)^{\frac{1}{2}} = (0.00098)^{\frac{1}{2}} = 0.031$$

EXAMPLE 2. Evaluate the data given in the following MISS PEACH cartoon.

MISS PEACH by Mel Lazarus. Courtesy Publishers-Hall Syndicate.

When told that Ira averages 2 baths per day, Mr. Grimmis assumed that the 14 baths were distributed evenly over the week. He was wrong, however, and we are amused at the actual distribution because it is so illogical. We know, from our own experience, that the data in the cartoon is ridiculous; is this also evident in the treatment of the data?

Day	Number of Baths	Deviation from Average	$\left(\dfrac{Deviation}{from\ Average}\right)^2$
Monday	0	2	4
Tuesday	0	2	4
Wednesday	0	2	4
Thursday	0	2	4
Friday	0	2	4
Saturday	14	12	144
Sunday	0	2	4

$\overline{X} = 14/7 = 2$ $\overline{d} = 24/7 = 3$

$$\sigma = \left(\dfrac{168}{8}\right)^{\frac{1}{2}} = (28)^{\frac{1}{2}} = 5.3$$

Notice that the average deviation is larger than the average. This is very strange to a chemist who is more accustomed to the average deviation being much smaller than the average as in Example 1. Ira's bath rate is not the type of measurement that chemists usually do so maybe this finding is not so strange after all. But we are dealing with a real system and things like the average deviation and the standard deviation should have real meaning if the treatment of data is valid. If one standard deviation is a reasonable error, then $\overline{X} \pm 1.0\sigma$ is 2 ± 5 or the range of baths per day is —3 to 7. Now the lower limit on the number of baths in any one day must be zero (how do you take minus one bath?). Thus the range cannot start at —3 but must start no lower than zero. The cartoon tells us that an average value isn't always the whole story. The treatment of data not only confirms this but reminds us that we must use our knowledge of the system under study to evaluate our data. Statistical analysis of data can often yield valuable information but it is always necessary to check to be certain that the analysis is in keeping with the physical meaning of the data.

When to discard a result. In the course of an experiment it is sometimes obvious when a result is going to be in error. Something has gone wrong: you spilled some solution or you overran the end point of a titration or your product should be white but it is still blue. Most of the time everything appears to have been satisfactory or there may be only a small doubt in your mind. It is in such cases that we need some guidance.

If there are a large number of measurements, an evaluation of a measurement that seems to be in error can be made using Table 3. If the deviation of the questionable measurement from the mean is more than twice the standard deviation, we can be more than 95% confident that it is in error. If it is only one standard deviation from the mean, then we can be less confident (68%) that it is in error and the measurement should probably be retained.

In most instances a chemist takes only a few measurements. Then it becomes much more difficult to decide when a result might be in error. A few guidelines are available but to apply these it is necessary to have at least three measurements.

The $4 \times \overline{d}$ test. If the measurement in question deviates from the average by more than four times the average deviation calculated by excluding the measurement in question, the measurement may be safely discarded.

EXAMPLE 3. Four samples of a material were analyzed for magnesium content. The results obtained were as follows: 11.88, 11.75, 11.90, and 11.49%. Can the last value be considered to be in error?

```
        %              d
      11.88          0.04
      11.75          0.09
      11.90          0.06
    3)35.53        3)0.19
      11.84          0.06
```

The value in question has a deviation from the average of 0.35. Four times the average deviation is (4) (0.06) = 0.24. Thus this value may be discarded.

The Q test. The rejection coefficient Q is another way by which a questionable value may be evaluated;

$$Q = \frac{v}{r}$$

where v equals the difference between the questionable value and the value closest to it and r equals the range of values (the difference between the largest and smallest values). Table 4 lists values of Q at two different confidence levels.[1] The questionable value should be rejected when the calculated Q exceeds the Q value from Table 4 at the desired confidence level.

TABLE 4

Values of Q for Q Test

	Number of Measurements				
	3	4	5	6	7
90% confidence level	0.89	0.68	0.56	0.48	0.43
95% confidence level	0.94	0.77	0.64	0.56	0.51

EXAMPLE 4. Apply the Q test to the magnesium analysis results given in the previous example.

$$v = 0.26$$
$$11.49 \quad 11.75 \quad 11.88 \quad 11.90$$
$$r = 0.41$$
$$Q = \frac{0.26}{0.41} = 0.63$$

If this value of Q, 0.63 is compared with the value of Q given in Table 4 for four samples at the 95% confidence level 0.77, it is seen that the observed value of Q is smaller and so the questionable value should be retained.

What's going on? The $4 \times \bar{d}$ test tells us to discard the value 11.49 but the Q test tells us to retain the value. What should we do? It is best to carry out more determinations. If we cannot do that, then we must resign ourselves to the fact that the tests we are using are not perfect. Indeed, the tests described here were chosen in part because of their simplicity and their applicability to small numbers of samples. Other more elaborate tests are available and may be found in tests.[2,3]

SPECIFIC REFERENCES

1. W. J. Dixon, *Ann. Math. Stat.*, **22**, 68 (1951).
2. K. Eckschlager, "Errors, Measurement and Results in Chemical Analysis", Van Nostrand Reinhold Co., London, 1969.
3. H. L. Youmans, "Statistics for Chemistry", Charles E. Merrill Publishing Co., Columbus, Ohio, 1973.

C
Balances

Balances are manufactured by many companies and have a large variety of features, varying capacities, and sensitivities. Since it is not practicable to cover all these possibilities, this section will consider some of the more common types of balances and emphasize their important working features and capabilities. The particular balance you will use may be slightly different in that knobs may be on the side rather than in front or have a slightly different readout system. If you are ever in doubt about how a particular balance operates, ask your instructor.

All types of balances will give good service only if they are given tender loving care. The two most important considerations are gentleness and cleanliness. Rough handling causes knife-edge fulcrums to become dulled and parts to be bent out of place. Chemicals are corrosive not only to the balance but also to the next person who uses the balance and who may not notice the spilled chemicals.

Since most balances work on the principle of the lever, it is necessary to level the beam under no-load (or constant-load) conditions before making any weighings. This is called *zeroing* the balance and the means by which it is done varies with balance type. Balances of low sensitivity (trip balances and triple-beam balances) generally do not need zeroing each time but balances of high sensitivity (some top loaders and analytical balances) need to be zeroed each time used.

Scientists always need to be concerned with the accuracy of their quantitative measurements. The limitations of each type of balance discussed here are given and those of any balance are usually obvious from the limitations with which the scales on the balance can be read. When using balances of low sensitivity, the accuracy of a weight measurement can often be improved by weighing the object or material by difference. When weighing solid chemicals or liquids, the most convenient way is to first weigh an empty container, then place the material to be weighed into the container, and finally obtain the weight by the difference (see Section D "Weighing of Samples" p. 23). When weighing other objects, the same procedure should be followed even though it seems like extra work.

Regardless of the type of balance used, these general procedures should be followed.

1. Check the pan or pans for cleanliness.
2. Zero the balance if necessary.
3. Carry out the weighing procedure with care and gentleness.
4. Arrest the beam if the balance has a beam arrest.
5. After weighing is completed, return all weights (and poises) to their zero positions.
6. Clean up.

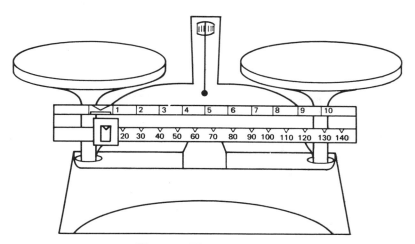

Figure 3. The trip balance.

1. TRIP BALANCE

The Harvard trip balance, with a sensitivity of 0.1 g, is a useful laboratory balance for rough weighings. Fig. 3 pictures a typical trip balance with two beams, that is, with two weight scales. The item to be weighed is placed on the left pan and the poises are slid along the beams until the pointer returns to the center of the scale. If the weight exceeds that of the poises, additional weights may be added to the right-hand pan in conjunction with the poises.

2. MULTIBEAM BALANCE

A typical multibeam balance is shown in Fig. 4(a) and a modification of that in Fig. 4(b) where one beam is replaced by a dial with vernier. These balances have a sensitivity of 0.01 g.

The general weighing procedure is as follows. With the pan clean and empty and all poises and dials set to zero, the pointer should read zero. If not, this needs to be set by means of knurled knobs near the left end of the beam; ask your instructor to do this for you.

Place the object to be weighed on the pan and slide the poises along the beams until the pointer returns to the zero on the scale. In the case of the balance with the dial, the final adjustment is made with the dial rather than with a poise. Fig. 5 shows close-up views of these balances. In both cases the weight shown is 2.35 g.

Older models of this type of balance need to be level. There is a bubble indicator and the balance is level when the bubble is in the center of the circle; it is adjusted by means of knobs on the front feet. Older models also have a beam-arrest mechanism. Except when actual weighings are being taken, the beam should be arrested; that is, the balance beam is raised off the knife edge, which protects it from damage.

3. TOP-LOADING BALANCE

Top-loading balances are simple to operate and weighing can be completed very quickly. Different models, even from the same manufacturer, may operate differently and so some specific instructions regarding the model you are to use may be necessary. The descriptions in this section should be sufficiently general to cover most aspects of all models.

Balances are manufactured from a capacity of 13,000 g with a sensitivity of 1 g to a capacity of 330 g with a sensitivity of 0.001 g (1 mg). The balance shown in Fig. 6(a) is one of the most sensitive; a typical readout is shown in Fig. 6(b).

The balance must be level; the bubble is adjusted to the center of the circle by means of knobs on the front feet. The light is turned on by means of the on-off switch (or by means

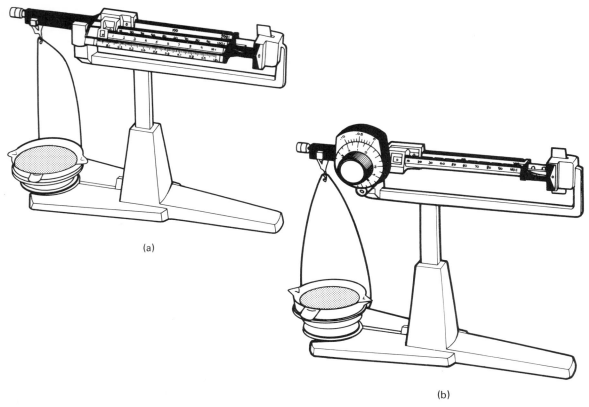

Figure 4. (a) The multibeam balance, (b) The Dial-O-Gram balance.

of a releasing knob on the left side). With all dials set at zero, the optical scale should also read zero. If not, it should be set to zero by lining up the line next to 0 with the fixed pointer (or by lining up the light area next to zero with the fixed line); the adjustment is made with a knob at the lower right-hand side of the balance. The object to be weighed is placed on the pan and the weight knob at the left is turned until the optical scale reads with numbers showing. Then the right-hand digital weight-setting knob is turned until the line next to a number lines up the fixed pointer (or until the light area next to a number lines up with the fixed line). The weight of the object is then read directly from the numbers showing in a horizontal line. Note that the decimal point is indicated by a comma rather than a period; this is because these balances are manufactured in Europe where the comma is customarily used as a decimal point. Fig. 6(b) shows a readout of 21.176 g.

These balances have a feature known as the *tare* that makes the weighing of samples into

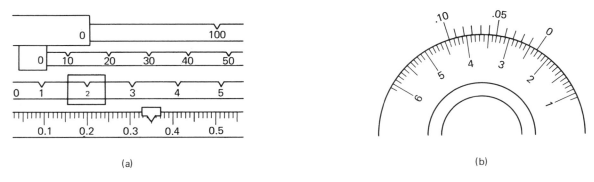

Figure 5. (a) Beams of a multibeam balance set at 2.35 g. (b) Dial of a Dial-O-Gram balance set at 2.35 g.

a container much easier and faster than with a multibeam or trip balance. Once the empty container has been placed on the pan, the optical scale may be set back to zero using the tare (*Tara* in German) knob. When the sample is placed in the container and the sample plus container is weighed, the numbers that appear on the readout will be the weight of the sample only.

4. ANALYTICAL BALANCE

The analytical balance, which can be read to 0.0001 g (0.1 mg), is one of the chemist's most important tools. It is an instrument that can provide information of high accuracy and precision, and it should be handled with care, thought, and attention to detail. A typical analytical balance is shown in Fig. 7(a) and a readout from this balance in Fig. 7(b).

This balance is referred to as a *single-pan*, *substitution-type* balance. It is completely mechanical in its operation except for an electric light bulb used to project a scale image onto a screen. It is called a substitution-type balance because an object is weighed by sub-

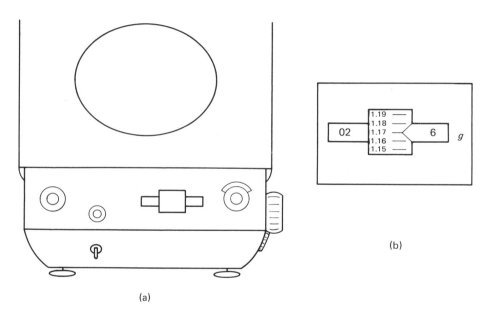

Figure 6. (a) A top-loading balance, (b) Readout of a top-loading balance showing 21.176 g.

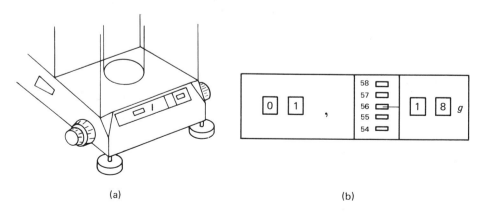

Figure 7. (a) An analytical balance, (b) Readout of an analytical balance showing 01.5618 g.

stituting the weight of an object placed on the pan for weights that are supported on a carriage just above the pan. Fig. 8 shows the construction of the balance schematically. Notice that the weights are on the same end of the beam as the pan; at the opposite end of the beam is a single large weight that just balances the weight of the weights, the weight carriage, and the pan. When an object is placed on the pan, the beam tilts (down on the left as shown in the diagram). In order to restore the beam to a balance position, weights must be removed from the weight carriage. Since only whole-gram weights are used, the fractional part of a gram is obtained by means of a small scale whose image is projected to the front of the balance by means of the light bulb and a series of mirrors. This arrangement of weights results in the balance always operating under the same total load, which means that the sensitivity does not change with the weight of the object being weighed as is true of double-pan balances.

The beam rests on a knife edge at its fulcrum. The continued accuracy of the balance is dependent on the knife edge remaining in good condition. Since vibrations, jarring, and other hard knocks might damage the knife edge, the balance mechanism is designed to lift the beam knife edge up in the air and hold it there firmly (this is called *having the beam arrested*). It is in this position that it is safe to put objects on the pan, take them off, clean the pan and balance case, and change weights. In order to assist in arriving at the point of making that final reading, the balance also has a position where the beam is set down on the knife edge but its freedom of motion is restricted. This allows gentle operations like adding or subtracting weights to be carried out with safety (this is called *having the beam partially released or half-released*).

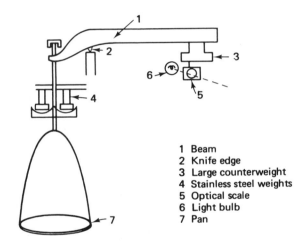

1 Beam
2 Knife edge
3 Large counterweight
4 Stainless steel weights
5 Optical scale
6 Light bulb
7 Pan

Figure 8. Sketch of internal parts of an analytical balance.

Precautions when Weighing

1. Fully release the beam only when adjusting the zero and when taking a final reading. At all other times the beam should be arrested.
2. Keep the balance clean.
3. Do not weigh chemical materials directly on the pan. Use a small container or a piece of glazed paper.
4. Do not handle objects being weighed with the fingers. Use tongs or a strip of clean paper.
5. Allow all objects and chemicals to attain room temperature before weighing; otherwise convection currents cause errors.
6. Perform all operations slowly and gently. The balance is a precision instrument and must be treated with utmost care in order for it to continue functioning to capacity.

Procedure for Weighing

1. Check that the beam is arrested; this is when the beam-arrest lever is in the center position.
2. If necessary, level the balance using the knobs on the front feet until the bubble in the indicator is in the center of the circle.
3. Brush the pan and inside the case to remove any dust, etc. Close the doors.
4. With all dials at zero and the pan empty, release the beam slowly to the full release position. Adjust the zero control knob so that the reading mark is at the 00 division of the optical scale. Arrest the beam.
5. Open the door. Place the object to be weighed on the pan. Close the door.
6. Slowly turn the beam-arrest lever to the partial release position. Turn the knob of the larger weights (units of tens of grams) until the optical scale swings back below zero; then turn the weight knob back by one. Turn the knob of the smaller weights (units of grams) until the scale swings back below zero; then turn the knob back one gram. Arrest the beam and pause before going on. Pre-weigh balances. Some balances have a feature that makes the weighing procedure a little faster. When the beam-arrest lever is moved in the direction opposite to that of full release, instead of partial release, the optical scale indicates the approximate weight of the object to the nearest gram. The weight dials can then be set immediately to that weight, the beam arrested, and the procedure continued with step 7.
7. Slowly turn the beam-arrest lever to the full release position. When the optical scale has come to rest, turn the optical scale control knob (NOT the zero control) until the reading mark is superimposed on the next division of the scale.
8. Read the results. Fig. 7(b) shows a typical readout with a weight of 1.5618 g.
9. Arrest the beam. Set the weights back to 00 and turn the optical scale control slowly back to 00.
10. Open the door and remove the sample. Brush off the pan and the case floor. Close the door.
11. Double check to see that the beam is arrested and that the balance case and pan are clean.
12. Report any difficulties to your instructor. And by all means, any time you have any doubt about what you are doing, ASK.

5. BUOYANCY CORRECTION

An object immersed in a fluid (liquid or gas) is buoyed up by a force equal to the weight of the fluid displaced. An object on a balance pan is buoyed up by the air. If the weights used were buoyed up to the same extent, there would be no error introduced. But the density of the weights is usually different from that of the object being weighed and consequently the buoyancy correction is not zero. If m_o is the true weight of an object, m_a is the weight in air, d_a is the density of air, d_o is the density of the object, and d_w is the density of the weights, then the relationship among these is

$$m_o = m_a + d_a \left(\frac{m_a}{d_o} - \frac{m_a}{d_w} \right)$$

At 20°C and 760 torr the density of air is 0.0012 g/ml and of stainless steel it is 7.76 g/ml.

When is it necessary to correct a weight for buoyancy? Let us consider two cases and calculate the correction for each.

Case 1. A 250 ml volumetric flask is being calibrated. Using stainless steel weights, the weight of water is found to be 248.3 g at 20.0°C. Calculate the true weight. The density of water at 20.0°C is 0.9982 g/ml.

$$m_o = 248.3 + 0.0012\left(\frac{248.3}{0.9982} - \frac{248.3}{7.76}\right)$$
$$= 248.3 + 0.26 = 248.6 \text{ g}$$

The error that would be introduced by neglecting the buoyancy correction is 0.3 parts in 248.6 or 1 part per thousand or 0.1%. If this much error can be tolerated, the buoyancy correction can be neglected; if not, then it must be made.

Notice that no correction is necessary for the buoyancy of the volumetric flask because the buoyancy of the flask is the same with or without the water and thus cancels out (along with the weight of the flask) when the weight of water is obtained by taking the difference of the weights of the full flask and the empty flask.

Case 2. A 1.0000 g sample of $BaSO_4$ is weighed, using stainless steel weights. Calculate the true weight; the density of $BaSO_4$ is 4.50 g/ml.

$$m_o = 1.0000 + 0.0012\left(\frac{1.0000}{4.50} - \frac{1.0000}{7.76}\right)$$
$$= 1.0000 + 0.00011 = 1.0001 \text{ g}$$

The error introduced in this case by neglecting the buoyancy correction is 0.0001 part in 1.0000 or 0.1 part per thousand or 0.01%. This is much less than in Case 1. Since the analytical balance is no more sensitive than 0.0001 g, the error introduced is no larger than the weighing error. Case 2 is very typical of the type of weighing that chemists often do and this calculation illustrates why they seldom apply buoyancy corrections.

D
Weighing Samples

1. METHODS OF WEIGHING

There are two basic methods that are used for weighing chemical samples and these can be used for solids or liquids on any kind of balance.

1. *Direct method.* In this method a portion of a chemical on a weighing paper or in a container is placed on the balance pan and weighed. Since chemicals should not be placed directly on the balance pan, a glazed paper, weighing pan, or other small container must be used. This container must be weighed or otherwise compensated for before (or after) weighing the sample. Aluminum or plastic weighing pans are very convenient and are reusable. Liquids should be weighed in flasks or beakers and volatile liquids must be weighed in stoppered containers.
   This method is faster than the difference method and when the weight does not need to be known very accurately or when moderate- to large-sized samples are being weighed, it is usually the method of choice. However, the direct method is equally applicable to the weighing of small samples on analytical balances.
2. *Difference method.* In this method the container, perhaps a weighing bottle, and the chemical that it is holding are weighed. A portion of the desired size that is to be used for an analysis, reaction, etc., is removed and the container and remaining chemical are reweighed. The sample weight is the difference between the two weighings. When using this method, one must be very careful not to spill any chemicals if the sample is to be used for precise quantitative work.
   This method is frequently used for analytical and quantitative experiments and for weighing small samples.

2. QUANTITATIVE TRANSFER OF SOLIDS AND SOLUTIONS

There are times when it is necessary to transfer a solid, liquid, or solution from one container to another without losing or leaving behind any of it. Some ways of accomplishing this are described in this section but you may be able to think of other methods, the only criterion being that none of the chemical may be lost or left behind.

Solids. Probably the only truly quantitative way to transfer a solid sample while maintaining it in a dry, solid form is to weigh it, either by the direct or difference method, into the container in which it is to be used. If the transfer does not need to be 100% complete, it is sometimes possible to weigh the sample on a glazed weighing paper or a weighing pan and then to carefully pour it into the desired container. Caution must be used as some chemicals will stick to weighing pans and papers and powdery chemicals may develop a static charge causing them to stick also.

The transferring of a solid sample that will be used to make up a solution is relatively simple.

> EXAMPLE 1. A solid sample is to be dissolved in a beaker and analyzed. The sample might be weighed in a weighing pan and then carefully poured into the beaker in which it will be analyzed. Some of the solvent that will be used to dissolve the sample is used to rinse the weighing pan, with the rinse solvent being caught in the beaker. The remainder of the solvent is added to the beaker and the sample is dissolved.

> EXAMPLE 2. A solid sample is to be dissolved to make up a solution in a volumetric flask. Again the sample might be weighed in a weighing pan. A funnel (preferably short-stemmed) is placed in the neck of the volumetric flask. The sample is poured into the funnel and some of the solvent to be used to make up the solution is used to wash the sample into the flask. The weighing pan is rinsed with solvent and the rinse poured into the flask. The funnel and outside of the funnel stem are rinsed thoroughly with solvent, with the rinsings going directly into the flask. Finally the neck of the flask should be washed down with solvent and the solution made up as usual. If the solid is insoluble or if it is not known whether or not the solid is soluble, the solid should first be dissolved in a beaker as in Example 1 above and then the solution transferred to the volumetric flask as in Example 6.

Sometimes it becomes necessary to transfer a precipitate that is suspended in a solvent or a solution. This procedure occurs frequently in a gravimetric analysis.

> EXAMPLE 3. A precipitate of AgCl in an aqueous solution is to be transferred from a beaker to a filter crucible so that it can be filtered, dried, and weighed. A stirring rod is laid across the top of the beaker and the solution and the AgCl are carefully poured off the end of the stirring rod into the filter crucible as shown in Fig. 9. The beaker and stirring rod are held in one hand while the other hand

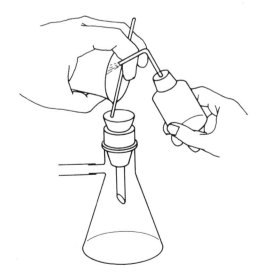

Figure 9. Transfer of a precipitate.

holds a water wash bottle that is used to wash down the inside walls of the beaker. A rubber policeman that is fitted on one end of the stirring rod is used to dislodge AgCl adhering to the surfaces of the beaker. Washing the beaker and pouring the rinse water into the filter crucible are carried on simultaneously until the beaker is completely clean. Finally, rinse the stirring rod and rubber policeman one more time. This method of pouring may seem awkward at first but it prevents droplets of the mixture from forming on the lip of the beaker and running down the outside.

Liquids. Liquids are handled in essentially the same manner as solids.

Solutions. The transfer of a solution from one container to another is a frequently performed operation in the laboratory. When the transfer must be quantitative, it requires a little care.

EXAMPLE 4. Pipeting a solution is one manner in which a solution can be transferred quantitatively. In this case not all the solution is transferred but just a known portion of it.

EXAMPLE 5. A solution contained in a small beaker is to be transferred to a larger beaker. A stirring rod is laid across the top of the beaker and the solution is carefully poured off the end of the stirring rod as shown in Fig. 9. The procedure is the same as that given in Example 3, except that now there is no precipitate to worry about. The original beaker and the stirring rod should be rinsed thoroughly with the solvent and the rinse added to the solution. In this case the concentration of the solution is changed, but the total amount of dissolved sample is the same.

EXAMPLE 6. A solution contained in a beaker or flask is to be transferred to a volumetric flask. The procedure is about the same as in Example 2. The beaker that contained the solution and the stirring rod used as a pouring aid should be thoroughly rinsed with solvent.

3. DRYING SAMPLES

Commercial chemicals and compounds prepared in the laboratory frequently are moist with water absorbed from the air or with some solvent and often must be dried before they can be used. The three most common reasons for drying a chemical are to obtain an accurately weighed sample of pure chemical, to prevent possible decomposition of the chemical in the presence of the moisture or solvent, or to prevent interference of the moisture or solvent in a reaction to be performed. In the experiments in this manual the chemicals are dried either for use in an analytical procedure or for determining the yield of a reaction product. In an analytical procedure it is of utmost importance to have a dry sample for weighing because a small error in the weight of the sample caused by moisture may cause a considerable error in the final results.

There are many methods by which a sample can be dried. The one chosen depends on the properties of the particular sample. Drying in an oven at a particular temperature is usually the easiest method. The temperature chosen must be below the decomposition temperature of the substance or else all will be lost. It is wise to use the lowest temperature that will completely remove the water or solvent. For oven drying stable substances that contain adsorbed water, $110°C$ is generally used. Substances to be dried for analytical procedures are usually dried in a weighing bottle or other small container that is placed in a beaker with a loose fitting cover as a dust shield, as shown in Fig. 10. The sample should be dried for 1 to 2 hours and stored in a desiccator after drying. Larger amounts of material, such as the product from a synthesis, can be dried in an open container, e.g., beakers, evaporating dishes, etc. Do not place plastic containers in an oven unless you are certain they can withstand the temperature. Polypropylene can be heated to 120°C but polyethylene lab ware cannot. *A characteristic of a dry sample is that the crystals or particles of the sample do not stick to each other or to their container.*

Figure 10. Two types of weighing bottles in a beaker ready for ovendrying.

A desiccator is a container with an airtight lid and it holds a material capable of absorbing moisture or solvent vapor. Thus the air and samples inside the desiccator are kept dry. Many materials are available as desiccants but the ones most frequently used for moisture are anhydrous calcium chloride and anhydrous calcium sulfate. The desiccant is usually placed in the bottom of the desiccator and the sample containers are supported on a plate or piece of wire gauze placed above the desiccant. The desiccator is used to store chemicals, crucibles, and bottles that have already been dried. Only under certain conditions can a desiccator actually be used to dry a sample.

E
Cleaning of Glassware

Dirty glassware is one of the unpleasant facts of life in a chemical laboratory. It is essential, however, that the glassware be clean and, in most cases, dry at the beginning of an experiment. The best and psychologically least painful approach is to wash all dirty glassware at the end of the laboratory period so that it will be ready at the beginning of the next period. It is easier to clean a piece of equipment before the residues have dried or hardened and while you still remember what the residues are.

Washing of glassware. Cleansing powder, water, a brush, and some effort will clean a large percentage of dirty glassware. A small amount of powder can be placed in the glassware, followed by a little water, or a powder-water mixture can be prepared and added to the piece of glassware. Using the brush, thoroughly scrub all inside or dirty surfaces of the glassware. Do not be afraid to use a little elbow grease! Cleansing powders tend to form a film on the surface of the glass and, therefore, the glassware must be rinsed thoroughly with tap water. Finally it should be rinsed with a few milliliters of distilled water from a wash bottle. Water will drain in an unbroken film from a clean piece of glassware. In this way you can check to see if you have cleaned it satisfactorily.

When cleaning volumetric glassware; i.e., volumetric flasks, pipets, and burets, never use any agent that might scratch the glass and consequently change the calibration. Volumetric flasks and pipets can be cleaned by filling them with soap solution. Pipets should be rinsed by repeated fillings with water, using a pipet bulb as always. Special brushes are available for cleaning burets.

For the removal of stubborn dirt, a hot solution of a strong detergent works well. Commercial products are available under brand names such as CONTRAD 70, FL-70, Alconox, Liqui-Nox, and CHEM-SOLV. For the removal of very stubborn dirt, stronger measures are called for. An ultrasonic cleaner with hot detergent solution cleans exceptionally well. A chromic acid cleaning solution is useful but it is *very corrosive* and must be used with great care. It may or may not be available in your laboratory. Essentially it is a solution of chromium(VI) oxide, CrO_3, in concentrated sulfuric acid and it is a powerful oxidizing agent. Trace amounts of chromium ions are left on the glass and if they are not completely rinsed off, they may cause errors in experiments involving complexing agents. A similar cleaning solution, called Nochromix, contains no metal ions. To clean a piece of glassware with such a sulfuric acid-based cleaning solution, a few milliliters is placed in the glassware and the walls carefully coated with it by rotating the glassware, or the piece of equipment to be cleaned is filled with the cleaning solution and allowed to stand for a while. When finished, do not discard the cleaning solution but carefully return it to its storage container.

Drying of glassware. Many times it makes no difference whether a piece of glassware is wet or dry but sometimes it must be dry. The best and easiest way to dry it is to let it dry in air but of course this is not practicable when a piece of glassware must be used immediately. There are a number of alternative methods available; they are all faster than air-drying but some of them are still not very rapid. (1) Some, but not all, types of glassware can be flame-dried with a burner. (2) Occasionally laboratories have a special oven available for drying glassware. Do not place wet equipment in any oven that is being used for chemical purposes. (3) If it is necessary to dry a pipet, it is best done by attaching the pipet to the water aspirator line and *sucking* it dry. (4) Glassware can be dried rapidly by rinsing it with a few milliliters of acetone and then air-drying it. The experimental procedure will indicate when this method should be used and wash bottles containing acetone will be available in the laboratory. The acetone dissolves the water and the solvent is discarded. The remaining acetone evaporates rapidly due to its high vapor pressure at room temperature. (5) Compressed-air lines located in some laboratories are frequently contaminated with oil droplets so that it is unwise to use them for drying purposes unless a filter has been attached to remove the oil.

F
Liquid Measure

In handling and measuring the volume of a liquid, the type of apparatus or container used is determined by the quantity you have, what you are going to do with it, and how accurately you want to know the volume. In this section we want to tell you something about available containers and how to use them in the measurement of volume.

1. BEAKERS AND ERLENMEYER FLASKS

As currently manufactured, all have volume markings imprinted on them. These are very rough indicators and with a beaker you may be able to do as well by visual estimation; it is more difficult to do that with an Erlenmeyer flask due to the sloping sides. You should not use these containers for measuring volumes but only as a rough guide to the amount in the container.

2. GRADUATED CYLINDERS

These are designed for measuring volumes of liquids and thus are more accurate than the imprinted beakers and flasks that are designed to hold liquids. The graduated cylinders are

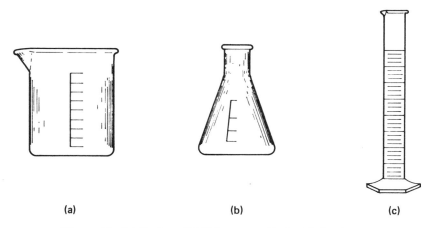

Figure 11. (a) Beaker, (b) Erlenmeyer flask, (c) Graduated cylinder.

manufactured in more than one grade (or quality). At worst they are accurate to ±1% and at best to ±0.4%. This means that if you measure out 40 ml in a 50 ml graduate, the error in your volume may be as large as 0.5 ml, just from errors in the calibration of the graduate. Other factors such as parallax* in reading and noncleanliness of the graduate may cause greater errors.

Most graduated cylinders are calibrated to deliver the amount indicated by the imprinted scale and such graduates will have the letters TD on them. The amount of liquid contained by the graduate should not be largely different from that which it delivers; at least it should be close to the tolerance levels. If this sort of difference is important to your experiment, however, then you are using the wrong type of equipment and you should consider using volumetric glassware as described below.

3. VOLUMETRIC GLASSWARE: FLASKS, PIPETS, AND BURETS

They are accurately calibrated liquid containers with each serving a different purpose. These containers are manufactured in a variety of sizes and in three general grades: Class A, which is the best; Class B, which is somewhat less accurate; and student grade, which is somewhat less accurate still. Class A has tolerances that do not exceed 0.2% of the indicated volumes; Class B has tolerances that do not exceed 0.4% of the indicated volumes; and student grade tolerances may be as large as 0.6% of the indicated volumes.

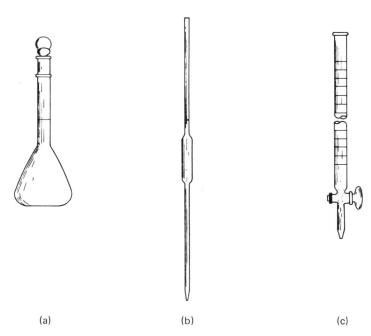

Figure 12. (a) Volumetric flask, (b) Pipet, (c) Buret.

Volumetric flasks are designed to contain (labeled TC) a known volume of liquid when filled to a designated mark. This type of container is convenient to use when preparing solutions whose concentrations are stated on a molarity basis since for such solutions you need to know the volume of solution.

To prepare a solution using a volumetric flask, a known weight of solute or a known volume of another solution is added to the flask. When adding liquids or solids, the use of a funnel is advisable since the neck of the flask is quite narrow. Then water is added to the

*The reading of liquid levels in glass cylinders will be found in the section on the buret.

flask until it is about two-thirds full (remember to rinse out the funnel *into* the flask) and the flask is then agitated by swirling until the solid is completely dissolved or the solution and the water are mixed. At this point, more water is added to the flask until the liquid is up into the narrow neck but still below the marking line. The last amount of water should be added with a dropper until the bottom of the meniscus is right on the line. This is an advisable procedure because the water added is like words out of your mouth; neither one can be called back. Once the proper amount of water is added, stopper the flask and complete the mixing by repeated inversions with swirlings and shakings in between. Keep your hand on the stopper at all times lest it fall out (or leak)!

Volumetric pipets are pieces of glass tubing with middle-age spread, calibrated to deliver (labeled TD) a fixed known volume of liquid when filled to the designated mark and allowed to drain. This type of measuring device is convenient to use when measuring a sample of solution where the volume must be known accurately but where the actual amount can be some predetermined fixed quantity (that quantity of course must coincide with an available size of pipet). With a little practice, dispensing volumes from pipets can be accomplished quickly and accurately.

In using a pipet, it is first necessary to rinse it with small portions of the liquid or solution to be used. Solution is drawn up into the pipet by suction and the pipet is tilted so that the solution runs throughout the pipet; then the solution is allowed to drain out into a container reserved for waste solution. This needs to be done two or three times. It is important to say a few words about the application of suction to draw liquid up into the pipet. People who are long on intrepidity due to shortness on perspicacity fill a pipet as if it were a straw. Since the pipet is not being filled with chocolate malted, however, some other form of suction is preferable to the mouth lest the mucous lining of the mouth be removed by base, fillings of the teeth be dissolved by acid, or illness ensue from ingestion of poisons (even seemingly innocuous chemicals may be poisonous). A rubber bulb is easy to use and completely avoids this problem.

Once the pipet has been rinsed out, squeeze the rubber bulb, attach it to the top of the pipet, place the tip of the pipet below the surface of the liquid, and slowly release the bulb until the liquid level is well above the mark on the upper portion (neck) of the pipet. Then remove the bulb and quickly cover the top of the pipet with the forefinger (NOT the thumb) before the level falls below the mark. While touching the tip of the pipet to the side of the container, release the pressure being applied by your forefinger slightly so that the level goes down toward the mark *very slowly*. Stop the flow out of the pipet several times before it reaches the mark so that you will get the feel of how much pressure to apply to stop the flow. When you have the liquid level exactly at the mark, touch the pipet tip to the side of the container once more and carry the pipet over to the container to which the sample is being transferred. Place the tip against the side of the container; check to see that the tip of

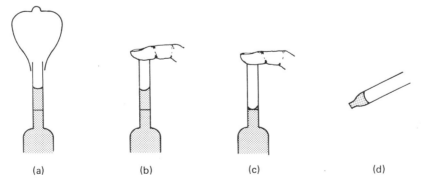

Figure 13. (a) A pipet bulb attached to a pipet with the liquid drawn up above the line, (b) Bulb removed and replaced by forefinger, (c) Liquid level adjusted to the line, (d) Liquid left in pipet tip after draining.

the pipet is full (an air bubble in the tip means you have lost some solution), and, if all is well, release your forefinger completely. Allow to drain; once drained, wait another 15 or 20 seconds to ensure complete drainage. There will be some liquid left in the tip of the pipet which SHOULD be there. DO NOT blow this out.

Data for the calibration of a 10 ml pipet are given in Table 5. Four aliquots of water were pipetted into a 50 ml flask and the ensemble weighed after each addition. The weight of water was corrected for buoyancy as described on p. 21, the volume of water delivered was determined from the weight of the water and the density of the water at the experimental temperature.

TABLE 5
Calibration of a Pipet

	Gross Weight (g)	Weight of H_2O (g)	Deviation from Average
Flask	36.078		
Flask + No. 1	46.045	9.967	0.002
Flask + No. 1, 2	56.018	9.973	0.004
Flask + No. 1, 2, 3	65.986	9.968	0.001
Flask + No. 1, 2, 3, 4	75.954	9.968	0.001
		4)39.876	4)0.008
		Average Weight = 9.969	0.002

Average Weight (g)	Buoyancy Correction (g)	Correct Weight (g)	Volume H_2O (ml)	Correction (ml)
9.97	0.01	9.98	10.01	+ 0.01

Temperature = 27.8°C
Density of water = 0.9964 g/ml
Density of weights = 7.8 g/ml

Burets are glass tubes designed and calibrated to deliver varying volumes of liquids. The two most common sizes are 25 and 50 ml with calibration marks such that it is possible to read volumes between 0 and 25 (or 50) ml with a reading estimated to ±0.01 ml. In the ensuing discussion we shall first consider how to take a reading of the level of liquid in a buret, then how to manipulate the stopcock to obtain desired volumes, and lastly how to design experiments using a buret including its calibration.

Fig. 14(a) shows a portion of a buret with the liquid level within the portion shown. Notice that the markings at each milliliter go completely around the buret and that each milliliter is divided into 10 equal parts; that is, each line is 0.1 ml. There are two precautions that must be followed in order to obtain precise (consistent) readings. The first is to reduce the light reflections so as to see a sharp meniscus of the liquid-air interface. This may be done by holding a white card behind the buret or, better still, a card such as that shown in Fig. 14(b). In this case, the dividing line on the card is to be lined up with the bottom of the meniscus. The second precaution is to avoid parallax errors. This is nothing more than assuring that your eye is on the same level as the bottom of the meniscus. Fig. 14(c) shows how, if you are higher or lower than the bottom of meniscus (which is in the center of the tube), the buret reading will be too high or low, respectively. If you watch the markings that go all the way around the buret, you can reduce parallax errors to a minimum.

Having taken the precautions described above, the buret reading is estimated to ±0.01 ml by mentally dividing the distance between two consecutive lines into 10 equal parts — remember that the lines have finite widths and so this distance should be from the top of one line to the top of the next. Thus the whole milliliters and the tenths are quite certain values but the accuracy of the hundreths is dependent on your carefulness and skill. The buret reading in Fig. 14(a) is 10.65 ml. *You* might look at this and say that the reading is 10.66 ml; this error is not large and the error in the volume used may be less if you are consistent in the manner in which you read the buret.

Data for the calibration of a student grade buret are given in Table 6. The weights were corrected for buoyancy and other calculations were done in a manner similar to that described previously for the calibration of a pipet.

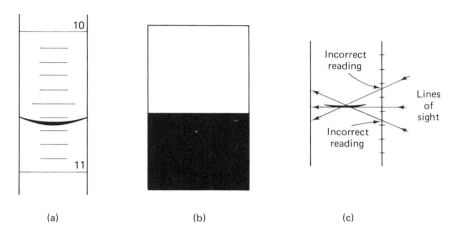

Figure 14. (a) Section of buret — see text for description, (b) Card to reduce reflections and aid in reading a buret, (c) Parallax.

TABLE 6

Calibration of a Buret

Buret Reading	Weight of Flask + H_2O (g)	Weight of H_2O (g)	Buoyancy Correction (g)	Corrected Weight of H_2O (g)	Volume of H_2O (ml)	Correction (ml)
0.00	37.631	0.000	0.00	0.00	0.00	0.00
5.00	42.570	4.939	0.01	4.95	4.96	−0.04
10.00	47.493	9.862	0.01	9.87	9.90	−0.10
15.00	52.473	14.842	0.02	14.86	14.90	−0.10
20.00	57.426	19.795	0.02	19.82	19.88	−0.12
25.00	62.414	24.783	0.03	24.81	24.88	−0.12
30.00	67.368	29.737	0.03	29.77	29.85	−0.15
35.00	72.354	34.723	0.04	34.76	34.86	−0.14
40.00	77.345	39.714	0.04	39.75	39.86	−0.14
45.00	82.284	44.653	0.05	44.70	44.83	−0.17
50.00	87.224	49.593	0.05	49.64	49.78	−0.22

Temperature = 24.2°C
Density of water = 0.9972 g/ml
Density of weights = 7.8 g/ml

4. STANDARD SOLUTIONS AND PRIMARY STANDARDS

A standard solution is one whose concentration is accurately known. Standard solutions are used extensively in quantitative analysis of materials and in experiments that require the amount of reagent to be known exactly. It is of extreme importance that standard solutions be prepared carefully, for the accuracy of any experiment involving them rests upon their accuracy of concentration.

Standard solutions may be prepared two different ways. A primary standard solution is prepared from a very pure solid compound, which is called a *primary standard substance*. To prepare such a solution, one would accurately weigh a sample of previously dried standard substance and dissolve it in water in a volumetric flask of the desired size as described on p. 30. The concentration of the solution is then easily calculated using the weight of the sample and the volume of the solution. Such a calculation would give either molarity or normality depending on whether the molecular weight or the equivalent weight of the primary standard is used. A secondary standard solution is a solution that is prepared and then standardized by reaction with a pure, solid substance. Some compounds are of insufficient purity or stability to be used as primary standard substances. In these cases, the compound is weighed approximately and dissolved in water in a volumetric flask so that the volume of the solution is known exactly. The concentration of the solution is determined by using it to titrate an accurately weighed sample of a suitable primary standard substance.

EXAMPLE 1. A solution of NaOH, approximately 0.1 M, is prepared by dissolving 4.0 g of NaOH in about 1 liter of water in a suitable flask or bottle. Some of this solution is placed in a 50 ml buret and is used to titrate a 0.8211 g sample of potassium hydrogen phthalate (KHP) that had been weighed into an Erlenmeyer flask and dissolved in 50 ml of distilled water. Phenolphthalein is used as an indicator; it turns from colorless to pink at the end point. The sample required 40.50 ml to reach the end point. What is the molarity of the NaOH solution?

The equation for the reaction is

$$OH^- + HC_8H_4O_4^- \longrightarrow C_8H_4O_4^{2-} + H_2O$$

According to this equation, 1 mole of OH^- reacts with 1 mole of $HC_8H_4O_4^-$, so the number of moles of NaOH used will equal the number of moles of potassium hydrogen phthalate in the sample.

$$\frac{0.8211 \text{ g}}{204.23 \text{ g/mol}} = 4.020 \times 10^{-3} \text{ mol KHP} = 4.020 \times 10^{-3} \text{ mol NaOH}$$

This amount of NaOH was contained in 40.50 ml (or 0.04050 l) of solution and so the concentration in moles per liter is

$$\frac{4.020 \times 10^{-3} \text{ mol NaOH}}{4.050 \times 10^{-2} \text{ l}} = 0.09926 \, M$$

In actual practice the titration would be repeated at least two times and an average value of the molarities taken as the concentration of the NaOH solution.

A solution could be standardized by titration with another standardized solution but this is not desirable because any error in the concentration of the first standardized solution will cause an error in the concentration of the second solution, aside from further titration or measuring errors.

Solutions, standardized or otherwise, should not be stored in volumetric flasks because it can be deleterious to the flasks. Standardized solutions should be stored in stoppered flasks or bottles. The bottle should be rinsed with two small portions of solution before transferring the bulk of the solution to it. The rinse solution is discarded. Substances that attack glass, such as NaOH and ethylenediaminetetraacetic acid solutions, should be stored in polyethylene containers.

5. USE OF VOLUMETRIC WARE IN EXPERIMENTS

Up to this point, the discussion has been centered on the mechanics of using volumetric glassware and on the maximum error introduced when these mechanics are properly performed. Now we need to look at the design of experiments to use this type of equipment most efficiently. If you are using a 50 ml buret, you should plan your quantities of reactants so that each sample will use about 40 ml of the solution in the buret. Why should one choose this volume? First, using the maximum volume allowed by the buret will minimize errors due to reading the volumes. If we say, as an example, that a typical error in volume due to errors in reading the buret is 0.02 ml, we can see that a volume of 10 ml will have a relative error of 2 parts per thousand (ppt); 20 ml, 1 ppt; 40 ml, 0.5 ppt. Thus the larger the volume, the smaller the error. But why not plan for using 50 ml? Since the experiment involves an unknown quantity, it is necessary to allow some room for a misjudgment in the unknown quantity. Usually in doing an experiment you have some idea of the approximate value of the unknown quantity, but if you don't, you make a wild guess and try it once. Then you will have enough information to plan the second sample well. The following example illustrates this.

EXAMPLE 2. There are a number of indicators available for a volumetric determination of chloride by titrating the chloride solution with a standard $AgNO_3$ solution. If a student has 0.1150 M $AgNO_3$ solution and a 50 ml buret available, what size sample should be weighed if the sample is approximately 30% chloride?

In 40 ml of the silver nitrate solution, there are 4.6×10^{-3} mole of silver ion that will react with 4.6×10^{-3} mole of chloride ion. This amount of choloride in grams is $(4.6 \times 10^{-3} \text{ mol})(35.45 \text{ g/mol}) = 0.16$ g. But the sample is only 30% chloride so to obtain 0.16 g of chloride would require 0.54 g of unknown sample. Thus for each titration, a sample of about 0.54 g needs to be weighed to the nearest 0.1 mg for four significant figures.

There is often more than one variable that can be adjusted to design a good experiment. This is illustrated by the following example.

EXAMPLE 3. A student has the following equipment available: 25 ml buret, 10 ml and 25 ml pipets, 100 ml and 250 ml volumetric flasks, and an analytical balance. He has a Na_2H_2EDTA solution that has been previously standardized and has a concentration of 0.01011 M. He wants to analyze a sample he has prepared for copper volumetrically and if it is what he thinks it is, it should contain about 23% copper.

First of all, let's calculate the size sample that would be required to react with 20 ml of this Na_2H_2EDTA solution (Cu^{2+} and $EDTA^{4-}$ form a 1:1 complex). Twenty milliliters of 0.01011 M solution contains 2.02×10^{-4} mole of EDTA and thus would react with 2.02×10^{-4} mole of copper. This amount of copper expressed in grams is $(2.02 \times 10^{-4} \text{ mol})(63.54 \text{ g/mol}) = 0.0128$ g. But the sample is only 23% copper so to obtain 0.0128 g of copper would require 0.056 g of sample. Now if we want to carry out the experiment to four significant figures, we need to weigh the sample to four significant figures. The minimum size sample we can weigh on an analytical balance to obtain this accuracy is 0.1 g. Weighing a sample of about 0.056 g would give only three significant figures and no matter how accurate the rest of the experiment, this one measurement is the least accurate and the final results can be no better than the least accurate measurement. So how do we get around this problem? We could weigh out 10 times as much, i.e., about 0.56 g, and prepare the solution in either the 100 ml or the 250 ml volumetric flask. Then we would take aliquots (10 ml from the 100 ml flask or 25 ml from the 250 ml flask) using the available pipets. Thus each aliquot would contain the desired amount and we would know this amount to four significant figures.

G
Filtration

Frequently during the course of an experiment, it becomes necessary to separate a solid from a liquid. The same techniques are used whether the purpose is to isolate a solid product or to remove a solid impurity or by-product from a liquid. The particular separation technique that is chosen is governed by the natures of the solid and liquid phases and the quantities to be separated.

1. DECANTATION

This is the simplest and fastest separation method but it can be used only when the separation need not be complete. Decant means *to pour off* and that is literally what is done in a decantation procedure. The solid phase is allowed to settle to the bottom of the beaker or flask. The liquid is carefully poured off, disturbing the solid as little as possible. This is done best by pouring the liquid down a stirring rod with the beaker held in one hand and the stirring rod held nearly vertical in contact with the lip of the beaker. It is not possible to remove all the liquid in this manner and therefore decantation is used when the solid is the desired product. It also makes washing of precipitates or reducing volumes of the liquid phase for subsequent filtration faster and easier.

It is frequently used when working with semimicro quantities of solids. In this case, the mixture is centrifuged to settle the solid phase before decanting the supernatant liquid.

2. GRAVITY FILTRATION

This is one of two general filtration techniques that are used to separate, essentially completely, a liquid from a solid. This technique utilizes the force of gravity to remove the liquid from the solid and is a rather slow process. It has the advantages of requiring very little equipment, merely a funnel and a piece of filter paper, and of being capable of separating a very finely divided precipitate from a liquid.

Filter paper comes in a variety of sizes, grades, and porosities. To filter a crystalline substance, a coarse filter paper should be used, while a finely divided precipitate would require a fine filter paper. For quantitative analyses, a specially washed grade of filter paper is available that can be burned without leaving an ash. In order to fit the funnel, the filter paper is (1) folded in half, (2) folded in half once more, and (3) a small corner is torn from the outside fold to ensure a better fit in the funnel. The filter paper should be of such a size that when folded it will almost reach the top edge of the funnel.

A long-stemmed funnel (15-20 cm) should be used for a gravity filtration, except under special circumstances. The funnel can be rested on the top of an Erlenmeyer flask of suitable size or can be supported from a ring stand by an iron ring or a clamp. The filter paper is placed in the funnel and moistened with a few drops of the same solvent that comprises the liquid phase to be filtered. Any air bubbles between the funnel and the paper should be removed because they will reduce the speed of the filtration. If the paper is properly fitted in the funnel, the stem of the funnel will remain completely filled with the liquid throughout the filtration. This fit is rather difficult to achieve and is not necessary if you are willing to sacrifice some efficiency. The filter paper in the funnel should be filled approximately three-fourths full with material to be filtered and should be kept near this level throughout the filtration. The more liquid in the funnel, the greater the hydrostatic pressure and the faster the filtration. On the other hand, if the filter paper is filled up to the top, precipitate will run down between the paper and the funnel and into your receiving flask, spoiling the filtration. The method for transferring a liquid and fine precipitate to a filter is discussed on p. 24.

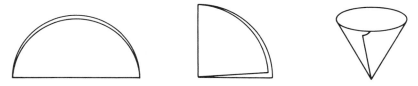

Figure 15. Folding a filter paper for filtration.

3. SUCTION FILTRATION

This is another filtration technique and it is used in preference to gravity filtration whenever possible. Suction filtration has the advantages of being rapid and of yielding a solid that is much drier.

The source of the suction (reduced air pressure) is an aspirator or water pump, which is an application of Bernoulli's equation.* The aspirator is a small attachment that fits on a water spigot and forces the water through a constricted portion of pipe. In order for the same volume of water per unit of time to flow through the constricted area as through the non-constricted area, it must flow faster through the constricted portion. The increased velocity produces a drop in the pressure and air is drawn in through a small hole in the side of the constriction. It is possible to obtain reduced pressures that are nearly as low as the vapor pressure of water at the temperature of the water.

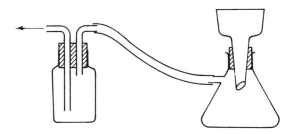

Figure 16. Apparatus for suction filtration including a trap.

*Bernoulli's equation relates the pressure, velocity, and elevation at points along a line of flow for an incompressible, nonviscous liquid.

A typical suction filtration apparatus is represented in Fig. 16. The trap is a very important part because it serves the purpose of preventing water from backing up into the filter flask and possibly contaminating the filtrate. At times water will come through the aspirator side arm if there is fluctuation in the water pressure in the line; i.e., when another person in the laboratory turns on an aspirator or if the water faucet is turned off before releasing the vacuum in the filtration system. Of course, if so much water backs up that it fills up the trap, it can then spill over into the filter flask but this is not a common occurrence. A gas wash bottle, another filter flask, or any heavy-walled flask can be used to make a trap. The longer piece of glass tubing should reach almost to the bottom of the trap but should not touch it. This long piece of tubing is attached to the aspirator by means of a piece of heavy-walled rubber tubing. The length of the glass tubing will permit water that may have backed up into the trap to be sucked out again when the water pressure in the pipeline has restabilized. The piece of glass tubing that is attached to the filter flask is made to extend into the trap for only 1 to 2 cm so that the trap must be nearly full of water before the water can flow into the filter flask. It is good practice always to break the suction in the system by removing a tubing connection before turning off the water faucet.

The filter flask is a heavy-walled Erlenmeyer flask that has a side arm attached to it. All rubber tubing connections in a suction filtration assembly must be of heavy-walled tubing that will not collapse under reduced pressure.

The most commonly used funnels and crucibles for suction filtration are illustrated in Fig. 17.

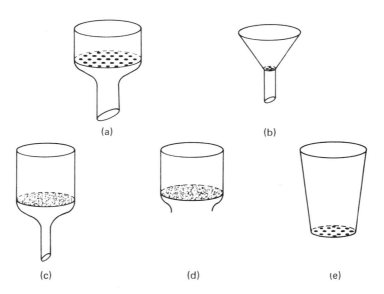

Figure 17. (a) Buchner funnel, (b) Hirsch funnel, (c) Fritted glass funnel, (d) Fritted glass crucible, (e) Gooch crucible.

(a) The Buchner funnel is made of porcelain and has a perforated plate that is covered with a piece of filter paper during a filtration. The filter paper should be moistened with a little solvent before beginning. The Buchner funnel is held in the filter flask by means of a rubber stopper. It comes in several sizes and is sometimes made of glass or polypropylene. It is not suitable for use in isolating very small quantities of a product, for quantitative work, or in cases where the chemicals will attack filter paper.
(b) The Hirsch funnel is similar to the Buchner funnel in that it is also made of porcelain and has a perforated plate. It is useful for filtering small quantities of solid materials, however, because of the small size of the plate. It is not suitable for quantitative work.
(c) The fritted glass funnel has a filter plate that is made of sintered glass and thus does not require the use of filter paper. It is made with frits having fine, medium, and coarse poros-

ities. It comes in a variety of sizes, including very small, and therefore is useful for isolating minute quantities of products. It has the disadvantage of being more difficult to clean than Buchner funnels or filter crucibles.

(d) The fritted glass crucible has a sintered glass filter plate but has no stem. It requires a device, such as the Walter's crucible holder, to hold it in the filter flask. This crucible is used extensively in quantitative analysis, generally for substances that can be dried at 110°C. The substance being analyzed is dried and weighed in a previously tared crucible.

(e) The Gooch crucible is made of porcelain and has a perforated filter plate. To retain the solid particles, a mat of shredded asbestos is placed on top of the filter plate by pouring an aqueous mixture of asbestos into the crucible and filtering off the water. If the asbestos mat is too thin, the precipitate will not be retained; if it is too thick, filtration will be too slow. To judge whether or not the thickness is correct, look through the crucible at a light. If the shadow of a pencil held against the bottom of the crucible is barely discernible, the thickness is correct. This system is good for removing a very fine impurity or by-product from a solution or liquid but it cannot be used in quantitative analysis in the same manner as the fritted glass crucible unless this crucible is tared with the asbestos mat in place.

Washing a precipitate during a suction filtration. After a precipitate has been isolated by filtration, it is frequently necessary to wash it with a solvent to remove adhering impurities. To facilitate this, the suction is broken by removing the tubing from the filter flask and the precipitate is allowed to remain in the funnel or crucible. A portion of the wash solvent is stirred (if you are using filter paper, be careful not to tear it; if you are using a Gooch crucible with an asbestos mat, do *not* stir it) and permitted to remain in contact with the solid for a few minutes before restoring the suction to the flask. This is repeated as many times as deemed necessary. The solid sample may be dried or partially dried by allowing air to be sucked through it for several minutes after the filtration has been completed.

H
Heating

Hardly a day in the laboratory will pass when some type of heat source will not be used for some purpose. There are a myriad of devices available for heating, each designed to fulfill a specific requirement, but all of them use either electric power or a burnable fuel as their energy source.

1. METHODS OF HEATING

The burner. Burners use a mixture of gas and oxygen (from air) as a combustion fuel and are a fast and efficient source of heat. They come in a variety of styles but all operate in basically the same manner. Three types of burners are pictured in Fig. 18. The gas flow is adjusted at the valve on the gas line in some types of burners. In others, there is a gas valve located on the bottom of the burner and in these types the gas line valve should be opened fully and the adjustment made at the burner valve. The airflow is regulated by adjusting the air intake valve, which is found at the base of the barrel. The gases are mixed as they flow up the barrel. Proper adjustment of the gas and air intake valves is necessary to obtain the desired type of flame. A hot flame requires a good amount of oxygen present in the gas mixture and it is characterized by a blue flame. A cooler flame results when the gas mixture contains only a small amount of oxygen and the hydrocarbon gas is incompletely burned. A cool flame is yellowish in color.

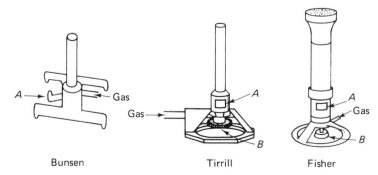

Figure 18. Types of burners. *A* is air adjustment; *B* is gas adjustment. Gas adjustment with the Bunsen burner is at the main valve.

The flame shown in Fig. 19 is the kind achieved with a properly adjusted burner. The center cone is blue and the region just above the surface of the cone is the hottest part of the flame. Objects that are to be heated strongly should be placed just above the top of the cone. The area inside the cone is cool because it contains fuel that is yet to be burned. Do not be hesitant to experiment with the adjustment of your burner so that you can obtain precisely the type of flame you need for a specific heating need.

The burner is lit by turning on the gas supply at the valve on the gas line and then lighting it with a match or striker. The flame is extinguished by turning off the gas. Be sure this valve is closed completely.

When heating beakers or flasks, always place a wire gauze between the container and the flame so that the heat is distributed over the entire bottom surface of the container. Do not heat flammable liquids with a burner.

Figure 19. Burner flame properly adjusted.

Other methods of heating. The electric hot plate is a convenient and frequently used source of heat. Although sometimes slow to heat up, once the desired temperature is achieved, it can be held constant indefinitely. There is much less danger of a fire with a hot plate as compared to a flame, although flammable liquids may ignite if they are spilled or boil over on a hot plate.

Heating baths consist of a container filled with some liquid and heated by a burner, hot plate, or built-in electric heating unit. These baths permit uniform heating over a large area of the heated flask. It is relatively easy to avoid overheating a substance by following the bath temperature with a thermometer and adjusting the heat source accordingly. It is a good way to heat flammable liquids providing a burner is *not* used as the heat source. For heating up to $100°C$, water is a good liquid for the heating bath. For higher temperatures, mineral oil can be used up to about $160°C$ and some silicone oils can be heated to $220°C$ or higher.

2. GLASS BENDING

It frequently becomes necessary in the course of a laboratory period to prepare a glass bend, fire-polish a stirring rod, or make capillary tubes or droppers. These operations are not so easy as they look but with a little practice most persons can master these simple manipulations.

There are two types of glass that are commonly encountered in the laboratory: borosilicate glass (hard glass) such as Pyrex or Kimax glass and soda-lime glass (soft glass). Borosilicate glass softens in the range of 800 to $900°C$, which is too high a temperature for it to be softened using an ordinary burner. It requires the use of a gas-oxygen torch. Most of your beakers and flasks will be made of this type of glass.

Soda-lime glass softens in the temperature range of 650 to $750°C$ and can be easily worked with a laboratory burner. Most of your glass tubing, stirring rods, droppers, and watch glasses will be made of soft glass.

When softening glass, you will need to use the hottest flame your burner produces. If you have a Bunsen-type burner, it is necessary to use a flame spreader in order to heat a large enough section of glass (see Fig. 22). If you have a Fisher-type burner, no attachment is necessary.

During the softening process the glass should be held in the hottest part of the flame. As the glass becomes hot, the flame around it turns yellow due to the excitation of sodium ions from the glass and their subsequent emission of yellow light.

Do not place hot glass on a cold bench top as it will be thermally shocked and may crack. Hot glass cools slowly and many burned fingers have resulted because people bending glass have forgotten this fact. So be cautious and carefully check the temperature of a piece of glass before touching it.

Cutting glass tubing. Place a piece of tubing on the lab bench and, using a triangular file or a glass scorer, make a single scratch on the tubing with a firm, stroking motion. Do not press down very hard, as you run the risk of smashing the glass and cutting yourself. A scratch produced by a single stroke is desirable because "sawing" on the glass frequently results in jagged breaks. If the scratch is not deep enough, however, the tubing will not break properly either and so it is permissible to scratch it a second time if you are careful to do it in exactly the same place as the first scratch. To break the tubing, hold it in both hands with the scratch facing away from you and place both thumbs on the tubing side opposite the scratch. Press firmly against the glass with your thumbs and at the same time pull your hands slightly apart. If all goes well, the tubing will snap cleanly in two pieces. If the cut ends are a little jagged, they can be smoothed off by rubbing them with a wire gauze. All cut ends of glass are very sharp and should be smoothed by fire-polishing.

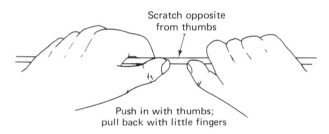

Figure 20. Breaking a piece of glass tubing.

Fire-polishing and preparation of stirring rods. Fire-polishing is the process of smoothing sharp edges of glass by heating them in a flame until the glass softens just enough to begin to flow. To fire-polish the freshly cut end of a piece of glass tubing, place the end in the flame and rotate the tubing continuously until the flame begins to glow yellow. Remove it from the flame and examine it visually; heat again if necessary. Prolonged heating will allow too much glass to flow toward the end of the tubing and the walls will thicken and eventually collapse inward.

A stirring rod is prepared by cutting a piece of solid glass rod and fire-polishing the ends.

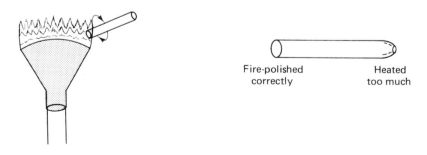

Figure 21. Fire-polishing the end of a piece of glass tubing.

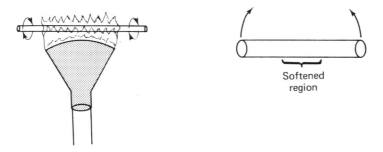

Figure 22. Bending a piece of glass tubing.

Glass bending. The preparation of a glass bend requires manipulation of the glass tubing while it is in a softened state. The preparation of a good bend will almost assuredly require a certain amount of practice on your part so don't be discouraged if your first attempts are less than you had hoped for.

Cut a piece of glass tubing about 20 to 25 cm long (unless there are restrictions on the length of the arms of the bend or on the lengths of tubing available to you). Hold the piece of tubing so that the center is in the hottest part of the flame and, with one hand on each end of the tubing, continuously rotate the tubing. The heated portion should be 3 to 5 cm in length in order to obtain a smooth, rounded bend. As the tubing softens and sags in the center, it is very important that the rate of rotation of both ends of the tubing be the same; otherwise the glass will be twisted in the center. Do not pull the ends apart or the glass will become thin in the softened region. When the glass becomes soft enough so that the ends of the tubing feel as if they are independent of each other, remove the tubing from the flame, hold it for 1 to 2 seconds (the time it takes to say "The University of Toledo"), and firmly and smoothly bring the ends of the tubing upward until a bend of the desired angle is obtained. After the bend has cooled, fire-polish the ends of the tubing.

A good bend will be smooth and rounded and the thickness of the glass walls in the bend will be uniform and approximately the same thickness as those in the unheated parts of the tubing. In a good bend, the inside diameter will be uniform over the whole length of the tubing; there will not be a constriction at the bend.

For instructions on the proper method of inserting a piece of glass tubing into a rubber stopper, see Section J-3.

Preparation of a capillary dropper. Sometimes it is necessary or convenient to have a dropper which is longer or which delivers smaller-sized drops than the droppers in your drawer. To make a dropper, obtain a piece of glass tubing and, holding it in a horizontal position just as you do when preparing a glass bend, rotate the tubing in the flame until the heated portion becomes very soft. Remove the tubing from the flame, hold it for 1 to 2 seconds, and move your hands apart with a smooth, constant motion. The softened tubing will stretch out into a long thin tube. The tubing can be cut off at the desired point with a file or glass scorer but this operation must be done carefully as the walls are quite thin and the tubing will crush or shatter easily.

The capillary tip will require *very little* heating in order to fire-polish it. The other end of the dropper may be just fire-polished but the rubber bulb will fit better if the end of the dropper is flared out slightly. This can be done by heating the end in a flame until it is quite

Figure 23. Preparation of a capillary dropper.

soft, removing it from the flame, placing the end of a file on the inside of the end of the dropper and rotating the dropper so that a lip is formed on the end.

The center portion of the pulled tubing can be used to make capillary melting point tubes. To seal off one end of the capillary, that end is rotated in a flame until the walls collapse and close.

3. HEATING OF LIQUIDS AND SOLUTIONS

Liquids and solutions are generally heated in flasks, beakers, and test tubes. In most cases the containers become quite hot so that they cannot be held with the bare hands. If they are to be heated with a burner, flasks and beakers are best supported on an iron ring attached to a ring stand, with a wire gauze between the flask and the burner flame. Test tube holders are available for holding a test tube while it is being heated.

When heating a solution in a test tube with a burner, the test tube should be inclined and a small flame should be used. Move the test tube around in the flame and agitate the contents gently to help distribute the heat throughout the solution. Never point the test tube at yourself or anyone else because the solution or liquid could "bump" and splash someone.

Bumping is the formation of a large vapor bubble at the bottom of the container that rises to the surface rather violently and it can cause some of the liquid to be forced out of the container. It occurs because the liquid becomes superheated at the point closest to the heat source. Bumping can be physically dangerous to persons and experimentally disasterous if quantative work is being done. There are several steps that can be taken to eliminate or alleviate bumping. If the solution or liquid is stirred or agitated, the heat will be more evenly distributed and superheating will be less likely to occur. A stirring rod placed in the liquid helps reduce bumping but boiling chips placed in the liquid are the most effective way to eliminate it. Boiling chips are pieces of material that have irregular surfaces to provide for the formation of many small gas bubbles. Commercially available boiling chips are inert to most chemicals except strong bases and concentrated sulfuric acid.

I
Instrumentation

1. MAGNETIC STIRRER

Even though it is always possible to stir a solution by using a stirring rod or by swirling the container, it is sometimes convenient to have a mechanical stirrer do it for us. Situations arise where both hands are needed for doing other things; since we don't have three hands, a motor can do part of the job for us. A mechanical stirrer is also helpful when there is apparatus in the solution such as a set of electrodes or when other apparatus is attached to the container holding the solution such as an addition funnel or a reflux condenser.

A magnetic stirrer has a magnet attached to a motor. The magnet rotates and its speed of rotation may be controlled by means of a knob (a rheostat) on the front. If another magnet, suitably encased in plastic or glass, is put in the solution and then placed on the magnet stirrer, as the stirrer magnet rotates, the magnet in the solution will rotate and thus stir the solution.

Some models have an on-off switch in addition to the speed control. There are other models that also have built-in heaters and thus can be used as a hot plate and a stirrer at the same time.

2. CENTRIFUGE

Once a solid is precipitated from solution, it can be separated from the supernatant solution by filtration as described in Section G. When the quantity of material is small (< 5 ml) and when the separation desired does not need to be complete, the method of centrifugation is rapid and easy. In doing qualitative analysis tests, sometimes it is difficult to tell the color of a precipitate when it is still dispersed throughout the solution. In such cases, centrifugation will pack the precipitate into the bottom of the tube where its color and quantity can be observed.

The operation of a centrifuge is simple. You put your test tube or centrifuge tube into one of the holes AND another test tube containing the same volume of water in the hole exactly opposite to provide a weight balance. Turn on the centrifuge for about 30 seconds. Once it has been turned off, you should let it coast to a stop unless it is equipped with a mechanical brake (do not brake too rapidly!). Remove your tube AND don't forget the tube with the water.

Centrifuges used for this purpose usually will take only 13 $\times$ 100 mm (10 ml) test tubes. If your test tube does not fit all the way into the holder so that only the lip is exposed, you are using too large a test tube. If the test tube used is too long, there is great danger of its breaking during the spinning.

Centrifuges are used in biology and biochemistry for separating materials that differ in density. Models come in a variety of sizes and enclosed in containers that can be heated or refrigerated. Quantitative studies, such as molecular weight determinations, can be done with an ultracentrifuge that can operate up to 70,000 revolutions per minute.

3. pH METER

One of the most common electronic instruments used by scientists is the pH meter. A pH meter is nothing more than a voltmeter especially designed to handle the high impedance of a glass electrode and to give a readout in pH instead of volts. Many models also can be used to measure voltages when the glass electrode is replaced by some other electrode.

When measuring pH, the electrode system used is a glass electrode whose potential is determined by the hydrogen ion concentration and a reference electrode whose potential is constant. Sometimes the two electrodes are both enclosed in the same glass container to give what is called a *combination* electrode.

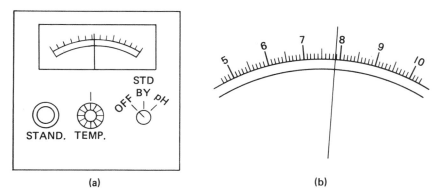

Figure 24. (a) pH meter, (b) Portion of meter reading a pH of 7.88.

A typical pH meter is shown in Fig. 24(a). All pH meters will have the three controls shown; some models may have one or two of these on the side and require the use of a screwdriver for adjustment. The function switch has three positions: one is OFF; the second is STANDBY or ZERO, which turns on the electronic circuit but does not include the electrodes in the circuit; and the third is READ or MEASURE or pH, which connects the electrodes into the circuit and is used for taking a measurement. The temperature knob should be set to the temperature of the solution being measured. The standardization knob is used to standardize or calibrate the dial as described below; once set, this knob should not be moved until the meter again requires standardization. Some models do not have an OFF position on the function switch; they are turned on by inserting the plug into an outlet and turned off by removing the plug.

Before use, it is necessary to standardize the pH meter with a solution of known pH. This is done using a commercial buffer solution or a "homemade" solution such as a saturated solution of potassium hydrogen tartrate (which has a pH of 3.56 at room temperature). Your instructor may already have standardized the pH meter; if not, this procedure is included in the procedure given next.

1 Procedure for Standardization and Use of pH Meter

1. Insert the plug into an outlet and turn the function switch to STANDBY. Allow the meter to warm up. Set the temperature dial to room temperature.

2. Rinse the electrodes with distilled water from your wash bottle.
3. Place the electrodes into the solution whose pH is to be measured or into a standard buffer solution if the meter is being standardized. The bulb of the glass electrode must be completely immersed as well as the junction of the reference electrode. Stir the solution carefully (by swirling) before taking a reading.
4. Turn the function switch to MEASURE.
   For standardization: Turn the standardization knob until the meter reading matches the pH of the buffer.
   For measurement: Read the meter dial. The meter can be read to two decimal places. Fig. 24(b) shows a typical meter reading 7.88.
5. Return the function switch to STANDBY.
6. Remove the electrodes from the solution; rinse them with distilled water. Replace the electrodes in distilled water as you found them.
7. If the pH meter is not to be used again during the laboratory period, turn it off.

CAUTION: Only turn the function switch to MEASURE when the electrodes are immersed in a solution. Except when actually taking a measurement, always turn the function switch to STANDBY.

CAUTION: Do not bump the electrodes. The glass electrode is especially fragile.

4. SPECTROPHOTOMETER

Even cursory examination of colored solutions reveals that the "darkness" of the solution, the intensity of the color, varies. Closer observation reveals three variables with respect to the color intensity. One is concentration of the colored solute. For example, a 0.1 M aqueous solution of $CoCl_2$ is light pink; a 1 M solution is darker pink — the more concentrated the solution, the more intense the color. The second variable is the amount of solution through which the light passes. For example, if the 1 M solution of $CoCl_2$ is used to fill a test tube, the solution will appear to be a darker pink when viewed through the end of the tube than when viewed through the side. Third, the nature of the solute affects the color intensity. For example, a 1 M solution of $MnCl_2$ is just barely pink, while a 1 M solution of $CoCl_2$ is quite pink. These observations are put together exactly in the Beer-Lambert law (commonly called *Beer's law*)

$$-\log \frac{I}{I_o} = A = abc$$

I_o = intensity of light directed at the solution

I = intensity of light being transmitted from the solution

a = absorptivity, usually in units of liter/mole · centimeter (formerly called extinction coefficient and sometimes denoted by ϵ)

b = path length, the thickness of the solution through which the light passes, usually in centimeters

c = concentration of the absorbing substance, usually in moles per liter of solution

A = absorbance (formerly called optical density, O.D.)

$\frac{I}{I_o} = T$ = transmittance

$100 \times T$ = percent transmittance

There is an additional variable that is not revealed by the preceding examples. They all used white light (and the eye as a detector); the absorbance is also affected by the wavelength of light used. This does not invalidate Beer's law. It means that the absorptivity value is wavelength dependent and the wavelength must always be specified for a particular value of the absorptivity.

The transmittance is the ratio of the intensity coming out of a solution (being transmitted from a solution) to the intensity of light entering. If the percent transmittance is 25, this means that 25% of the light is going through the solution and 75% is being absorbed by the solution. If no light is being absorbed, the percent transmittance is 100; if all the light is being absorbed, it is 0%. The absorbance A is related to the amount of light being absorbed. If no light is being absorbed ($\%T = 100\%$), then the absorbance is zero. The more light that is being absorbed, the larger the value of A; if all the light is being absorbed, A is infinitely large.

The particular experimental means by which spectral measurements are carried out depends on the accuracy required and the equipment available. Of course, the more accurate equipment costs more money. The simplest, which uses white light, is generally called a *colorimeter*. If the white light is passed through a filter to give a narrower range of wavelengths, the apparatus is often called a *photometer*. When a very narrow range of wavelengths is obtained by the use of a prism or a grating, the apparatus is called a *spectrophotometer*.

Colorimetry experiments may utilize equipment as simple as test tubes or more complex such as a Duboscq-type colorimeter. A solution of known concentration may be placed in a test tube to a measured height. Then the height of the unknown solution in a second test tube is varied until the intensity of the color appears to be the same as the first when viewed through the open end. Beer's law may then be applied where, since a and A are the same in both test tubes, the concentration must be inversely proportional to the height. The Duboscq-type colorimeter accomplishes the same thing by means of movable cells and mirrors. If the concentrations of the known and unknown solution are not close, then Beer's law may not be followed. The eye is not a very sensitive detector of light intensity, especially if the light being transmitted is not between 500 and 625 nm. If there is more than one colored species in the solution, colorimetry does not work at all.

Photometers are more accurate than colorimeters because a phototube (an electronic tube whose electrical current output is proportional to the light intensity impinging on it) is used to replace the eye as a detector and the range of wavelengths entering the solution is narrowed by the use of a filter. Such photometers, called *filter* photometers, usually have a set of filters, each one transmitting a different range of wavelengths. A typical wavelength range is 50 nm, which is considerably larger than that obtained from the use of a prism or a grating as done in spectrophotometers. Except for the means of choosing the range of wavelengths, the general operation of filter photometers is very much like that of a spectrophotometer.

Spectrophotometers use a phototube to detect the transmitted light, just as with filter photometers. In some of the less expensive models of spectrophotometers the range of wavelengths from the prism or grating entering the solution is 20 nm in the visible region of the spectrum. In more expensive models, this range is smaller. The phototube detector is connected to a meter that, by means of the electronic circuit, is read directly in absorbance or percent transmittance. The spectrophotometer described here is a simple one that operates in the visible region of the spectrum and is applicable to experiments in Part II.

Spectrophotometers are also manufactured that operate in the infrared region and others for the ultraviolet region of the spectrum. Except for the Beckman model DU (a visible-UV spectrophotometer), most infrared and ultraviolet instruments have graphical recorders and operate semiautomatically. Typical visible spectra are shown in Fig. 25-1 and 26-2 on pp. 218 and 225.

Typical spectrophotometers are shown in Fig. 25; a schematic drawing of the essential internal elements is shown in Fig. 26, and a meter in Fig. 27. All spectrophotometers will have the three basic controls shown. The wavelength selector rotates the prism or grating to direct the wavelength of light indicated on the dial to pass through the solution. The zero

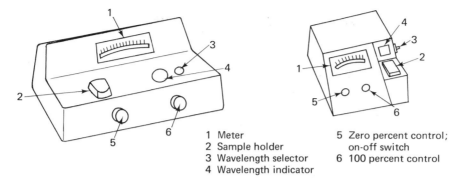

1 Meter
2 Sample holder
3 Wavelength selector
4 Wavelength indicator
5 Zero percent control; on-off switch
6 100 percent control

Figure 25. Spectrophotometers.

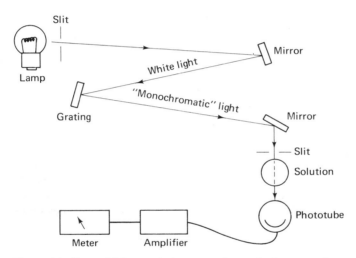

Figure 26. Essential internal elements of a typical spectrophotometer.

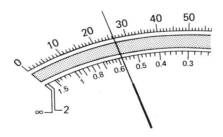

Figure 27. A spectrophotometer meter reading 26.3% T (or 0.580 absorbance units).

percent control is used to set the response of the phototube as indicated by the meter to 0.0% transmittance when no light is falling on the phototube; this darkness is produced by a shutter being placed in front of the phototube. The one hundred percent control is used to set the response of the phototube as indicated by the meter to 100.0% transmittance when no light is being absorbed by the solution; this is produced by inserting water or the solvent into the light path. The 100% setting is dependent on the wavelength of the light because the sensitivity of the phototube varies with wavelength.

II. Procedure for Use of Spectrophotometer

1. Insert the plug into an outlet and turn the spectrophotometer on. Allow the meter to warm up; generally 15 minutes is required.
2. Turn the wavelength control until the desired wavelength is indicated on the dial.
3. With no test tube or cuvette in the sample holder and the sample holder cover closed, set the meter to 0% transmittance using the 0% control.
4. Clean a test tube or cuvette, fill it at least half full of distilled water or solvent, and wipe off the outside with a lintless towel.
5. Insert the test tube or cuvette into the sample holder and close the cover. (Be sure to align the special mark if using a selected test tube or, if using a regular test tube, choose a mark on the test tube for alignment purposes.)
6. Adjust the 100% control until the meter reads 100% transmittance.
7. Repeat steps 3, 5, and 6 until both check with no adjustment of the knobs required.
8. Fill another clean test tube or cuvette at least half full with the solution to be measured. Wipe off the outside with a lintless towel.
9. Remove the reference tube and insert the one containing the sample. Close the cover and read the percent transmittance or the absorbance directly from the meter. Generally percent transmittance should be read because it is a linear scale on the meter and can be read with greater precision.
10. Remove the tube or curvette from the sample holder. Turn off the instrument if your measurement is the last of the day.

Remember that the light beam passes through the lower part of the test tube or cuvette. Thus it is important that this part be clean and free of dirt including *fingerprints*, so handle the tubes by the upper portion.

J
Mélange

1. READING A VERNIER

When taking a numerical reading from a piece of apparatus, the means of obtaining the last digit varies. Manufacturers have developed a digital readout system such as illustrated on p. 19. The other most common means is a vernier scale. A vernier is a scale, usually fixed in place, adjacent to another scale. It reads from 0 to 10 and provides a means of reading the last digit. Shown here is a scale with adjacent vernier.

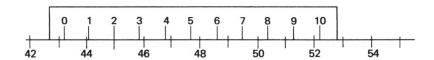

The zero of the vernier is between the 43 and 44 on the scale so the reading is 43.X. To decide on X, "eyeballing" the scale reveals that the zero on the vernier is close to the 43 line so X will be 1, 2, or 3. To tell exactly, check the lines on the vernier; the vernier line that exactly lines up with a line on the regular scale gives that digit. In this case, the last digit is 2 so the reading is 43.2.

The Dial-O-Gram balance has a vernier scale; see Fig. 5(b) on p. 18.

2. USE OF A WASH BOTTLE

Just in case you haven't become aware of it, this note is included here to point out the "handy-dandy" wash bottle. It is unexcelled for dispensing distilled water: for the final rinsing of glassware after washing; for adding distilled water to graduated cylinders, beakers, etc., when needed; for rinsing out pipets, burets, and volumetric flasks; for transferring precipitates; and for other routine jobs requiring small quantities of distilled water. If such things are done directly under the distilled water tap, a lot of distilled water is wasted, time is wasted because the distilled water tap becomes tied up, and your instructor's temper starts to flare. So to keep your instructor happy and to be more efficient, keep some distilled water in your wash bottle for handy use.

3. INSERTING TUBING INTO A RUBBER STOPPER

One of the most common laboratory accidents is a cut hand received from a piece of glass tubing or a thermometer that has snapped while being inserted into a rubber stopper. A few precautions can help you avoid this accident.

First, be certain that the hole in the stopper is large enough to accommodate the tubing or thermometer and still give a snug fit. If the hole is not large enough, enlarge it, use a cork stopper, or look for another way to set up the apparatus. Second, lubricate the tubing and stopper with glycerine or water to permit the tubing to slide more easily. Third, hold the tubing near the stopper, not out near the end. If the tubing has a bend in it, do not put pressure on the bend since it will break rather easily under pressure. Holding the tubing near the stopper minimizes the effect of any sideway force that may be inadvertently applied to the tubing. Fourth, wrap a towel around the hand (including the fingers) holding the stopper. If you can manage it, it is also a good idea to hold the piece of tubing with another towel although this makes it difficult to get a good grip on the tubing. Last, rotate the tubing as you push it into the stopper. This helps to prevent the tubing from sticking to the rubber as it slides through the hole.

4. INTERNATIONAL SYSTEM OF UNITS

In 1960, the international authority on units, the General Conference of Weights and Measures, agreed to establish an international system of units called SI Units (the abbreviation SI coming from the French Système International d'Unités). The SI Units have been modified slightly since then and have been incorporated by the International Union of Pure and Applied Chemistry into its *Manual of Symbols and Terminology for Physiochemical Quantities and Units.*[1] Since the system is based on the meter, the kilogram, the second, and the ampere, it is not in this regard a new system. What is new about it is that it discourages scientists from using any units other than SI Units.

In the Système International each distinct physical quantity has only one unit. That unit is either a base unit (Table 7) or is derived from base units (Table 8). For convenience, it is permissible to use certain multiples of the SI Units using the prefixes listed in Table 9. There are other physical quantities that are derived from SI Units and Table 10 lists some (but by no means all) of them. Other units which are commonly used but which do not fit into this prescribed pattern are listed in Table 11; the General Conference hopes to gradually phase these units out. Indeed, this process is already evident in the chemical literature in that the unit millimicron, $m\mu$, is being replaced by the nanometer, nm.

In this book, we have chosen to continue to use the traditional unit of energy, the calorie. We do this not because of any fond attachment to the calorie but because it is convenient. Also, there is enough past work written in traditional units that it is necessary for scientists to be familiar with both traditional and SI Units. It seems unlikely that chemists will readily accept the SI unit of concentration of mole per cubic meter, mol/m^3. The traditional concentration unit of molarity gives numbers that lie in a range (generally 0.1 to 10) convenient to use. These numbers in moles per cubic meter would be 0.0001 to 0.01, which are just not as convenient to work with. (Notice that the unit of mole per liter is the same as mole per cubic decimeter, mol/dm^3.)

SPECIFIC REFERENCE

1. "Manual of Symbols and Terminology for Physicochemical Quantities and Units", *Pure Appl. Chem.*, **21**, 1 (1970). See especially part 3.

GENERAL REFERENCES

M. A. Paul, "International System of Units (SI)", *Chemistry*, **45** (9), 14 (1972).
G. Socrates, *J. Chem. Educ.*, **46**, 710 (1969).
National Bureau of Standards, "Policy for NBS Usage of SI Units", *J. Chem. Educ.*, **48**, 569 (1971).

TABLE 7
SI Base Units

Physical Quantity	Symbol Quantity	Name of SI Unit	Symbol for SI Unit
Length	l	meter	m
Mass	m	kilogram	kg
Time	t	second	s
Thermodynamic temperature	T	kelvin	K
Amount of substance	n	mole	mol
Electric current	I	ampere	A
Luminous intensity	I_v	candela	cd

TABLE 8
SI Derived Units

Physical Quantity	Name of SI Unit	Symbol for SI Unit	Definition
Energy	joule	J	$kg\ m^2\ s^{-2}$
Force	newton	N	$kg\ m\ s^{-2}$
Pressure	pascal	Pa	$kg\ m^{-1}\ s^{-1}$ = $N\ m^{-2}$
Electric potential difference	volt	V	$kg\ m^2\ s^{-3}\ A^{-1}$ = $J\ A^{-1}\ s^{-1}$
Electric charge	coulomb	C	$A\ s$

TABLE 9
SI Prefixes

Fraction	Prefix	Symbol	Multiple	Prefix	Symbol
10^{-1}	deci	d	10	deca	da
10^{-2}	centi	c	10^2	hecto	h
10^{-3}	milli	m	10^3	kilo	k
10^{-6}	micro	μ	10^6	mega	M
10^{-9}	nano	n	10^9	giga	G
10^{-12}	pico	p	10^{12}	tera	T
10^{-15}	femto	f			
10^{-18}	atto	a			

TABLE 10
Other SI Derived Units

Physical Quantity	Name for SI Unit	Symbol for SI Unit
Area	square meter	m^2
Volume	cubic meter	m^3
Density	kilogram per cubic meter	$kg\ m^{-3}$
Velocity	meter per second	$m\ s^{-1}$
Pressure	newton per square meter	$N\ m^{-2}$
Concentration	mole per cubic meter	$mol\ m^{-3}$

TABLE 11

Non-SI Units with Special Names

Physical Quantity	Name of Unit	Symbol for Unit	Definition
Time	minute	min	60 s
	hour	h	3600 s
	day	d	86,400 s
Volume	liter	l	10^{-3} m^3
Length	ångström	Å	10^{-10} m
	micron	μ	10^{-6} m = μm
	millimicron	mμ	10^{-9} m = nm
Energy	erg	erg	10^{-7} J
Concentration	mole per liter	M	10^3 mol m^{-3}

5. COMPUTERS

The digital computer has enabled scientists to do calculations that never could have been done manually. But computers are also useful for simpler calculations, especially if the calculation to be done would be tedious by hand. The recent advent of time-sharing computers has made the use of a computer available to large numbers of students.

There are several computer programs included in this manual. If you have taken a computer course, then you should have little difficulty adapting these programs to the facilities available to you. The programs are in the language BASIC PLUS and have been used at The University of Toledo on a Digital Equipment Corporation's PDP-11 system. If you have never taken a computer course, do not be disheartened because it is not too difficult to learn how to write computer programs. You will need some advice from your instructor as to what computer facilities are available, what language is to be used, and where to find instructional materials. The general references listed below include a series of articles from *Chemistry* magazine that was written for students like you. The first two articles in the series are general and the third considers programming in FORTRAN.

In the programs included in the Appendix of this manual, the initial statements that begin REM are only comments about the program and are not part of the computational portion. The programs have been written to provide clarity of input data and printed results by using lots of space and tabular format. The results are rounded off to an appropriate number of significant figures. In the programs involving the determination of slopes of lines, it is always good practice to plot the data by hand in order to visually check for any point or points that deviate significantly from the line or if the data do indeed give a straight line. The printing out of intermediate results helps you in doing this. In addition, an evaluation of each point is printed out in terms of the deviation of the actual point from the calculated point.

GENERAL REFERENCES

D. W. Edman, M. M. Cox, and J. W. Moore
"Computers and Chemistry" I. *Chemistry*, **45** (1), 6 (1972).
II. "Algorithms and Problem Solving", *Chemistry*, **45** (3), 10 (1972).
III. "FORTRAN Programming", *Chemistry*, **45** (5), 10 (1972).
IV. "Chemical Application of Digital Computers", *Chemistry*, **45** (8), 13 (1972).

T. R. Dickson, "The Computer and Chemistry", W. H. Freeman and Co., San Francisco, Calif., 1968.

T. F. Fry, "Computer Appreciation", Butterworths, London, 1970.

T. L. Isenhour and P. C. Jurs, "Introduction to Computer Programming for Chemists", Allyn and Bacon, Inc., Boston, Mass., 1973 (FORTRAN version), 1974 (BASIC version).

Part II
EXPERIMENTS

Experiment 1

Measuring Volume by Counting Drops

When working with small amounts of material, it is often convenient to be able to estimate the volume of a liquid or solution being added by counting the number of drops from a dropper. For water and dilute aqueous solutions it is often assumed that a medicine dropper delivers 1 milliliter in 15 to 20 drops. In this experiment we shall check this out. In part A, this check is made very roughly using a graduated cylinder. Ethanol is also used to see what difference it makes when this technique is used for a liquid other than water. Then in part B, a more careful evaluation is made using water. Doing a careful evaluation will provide an opportunity of see how reliable such a volume measurement is and an opportunity to learn about treatment of data as related to multiple measurements. You will find a discussion of data treatment in section B-4 of Part I and a discussion of graduated cylinders and the reading of a meniscus on pages 29 and 33 which you should read before doing this experiment.

PROCEDURE

A. Add distilled water dropwise from a medicine dropper into a 10 ml graduated cylinder. Count the drops necessary to bring the level to the 5 ml mark. You will need to refill the dropper several times. After each time the dropper is filled, touch the dropper to the side of the beaker containing the water to remove any excess water from the tip. Calculate the number of drops per milliliter.

Empty the graduated cylinder and dry it out with a towel. Repeat the procedure using 95% ethanol in place of the water. If the dropper is not dry, fill it once with the ethanol (discarding this) before beginning the addition to the graduated cylinder. Calculate the number of drops per milliliter.

B. Weigh to the nearest 0.1 mg a clean container that will hold 25 to 50 ml (beaker or Erlenmeyer flask). From a clean medicine dropper, add 10 drops of distilled H_2O and reweigh. (How are you going to pick up the beaker or flask to put it on and take it off the balance? If you are not certain of the answer, see p. 20). Repeat with 10 drop increments until you have at least 12 measurements.

Measure the temperature of the water.

CALCULATIONS

From your measurements in Part B, calculate the weight of water delivered for each increment. Convert each weight to volume using the density of water at the appropriate temperature from Table 2-1 of Experiment 2.

Calculate the average volume per 10 drops and the absolute value of the deviation of each measurement for this average. Square each of these deviations, record the result to five decimal places, and calculate the standard deviation. Evaluate your results in light of this standard deviation and the probability table on p. 12.

Compare your results in part A with water and with ethanol to see the difference between two substances with different surface tensions. In part B, look at the deviations from the average, looking particularly at the range of deviations. From your results, decide how good a practice it would be to measure volumes of water or dilute aqueous solutions by counting drops.

EXERCISES

1. At a temperature of 23.0°C, a sample of water was found to weigh 1.0058 g. What is the volume of this water sample?
2. For the following set of numbers, calculate (a) average, (b) average deviation, and (c) standard deviation.

    55.54, 53.96, 54.42, 56.23, 53.87, 53.67, 55.06 57.75, 51.86

3. The average of a series of measurements was calculated to be 33.55 with a standard deviation of 3.30. In what range could you be 87% confident of finding a single measurement?
4. A student determined the following percentages of copper in a sample in four separate determinations: 25.33, 25.43, 25.06, and 27.01. The last measurement seems much larger than the others. Should it be discarded?

(Answers are on p. 241)

REPORT MEASURING VOLUME BY COUNTING DROPS

Part A.

1. Number of drops of water in 5 ml _____
2. Number of drops of water per milliliter _____
3. Number of drops of 95% ethanol in 5 ml _____
4. Number of drops of 95% ethanol per milliliter _____
5. Water temperature (°C) _____   Density of water _____

Part B.

6. No. drops in flask	Flask + water (g)	Increment of 10 drops (g)	Volume of increment (ml)	Deviation from average	$\left(\dfrac{\text{Deviation}}{\text{from average}}\right)^2$
0	_____	_____	_____	_____	_____
10	_____	_____	_____	_____	_____
20	_____	_____	_____	_____	_____
30	_____	_____	_____	_____	_____
40	_____	_____	_____	_____	_____
50	_____	_____	_____	_____	_____
60	_____	_____	_____	_____	_____
70	_____	_____	_____	_____	_____
80	_____	_____	_____	_____	_____
90	_____	_____	_____	_____	_____
100	_____	_____	_____	_____	_____
110	_____	_____	_____	_____	_____
120	_____	_____	_____	_____	_____
130	_____	_____	_____	_____	_____
140	_____	_____	_____	_____	_____
150	_____	_____	_____	_____	_____
Averages:					
10	_____	_____	_____		_____

7. Calculate the standard deviation, σ.

8. Using your results, complete the following statements.

    (a) There is a 68% probability that a 10-drop increment will have a volume between _____ ml and _____ ml.

(b) There is a 95% probability that a 10-drop increment will have a volume between _____ ml and _____ ml.

9. How many of your 10-drop increments are outside the 95% confidence limits?

10. In Part A, it would be faster to count the number of drops needed to fill the graduated cylinder to the 1 ml mark instead of the 5 ml mark. Why is this not done?

Name _____ Section _____ Grade _____

Experiment 2

Calibration of Volumetric Glassware

The use of volumetric glassware, flasks, pipets, and burets has been discussed in Part I under *Liquid Measure*. Included there was an estimation of errors involved in the use of volumetric glassware due to the manufacturing process. It is possible to reduce these errors by calibrating the glassware before it is used in an experiment. The calibration makes the measurements taken closer to the true values and thus one source of error in the experiment is reduced (assuming, of course, that errors are not made in the calibration). The data and results of the calibrations of a pipet and of a buret are given in Part I and you should refer to *Liquid Measure* for instruction on how to use volumetric glassware and on how to calculate corrections for calibrated glassware.

The procedure described in this experiment was used in obtaining the data given in *Liquid Measure*, thus you can use the information there as a model for your experiment.

TABLE 2-1

Density of Water

Temperature (°C)	Density (g/ml)	Temperature (°C)	Density (g/ml)
20.0	0.9982	25.0	0.9970
20.5	0.9981	25.5	0.9969
21.0	0.9980	26.0	0.9968
21.5	0.9979	26.5	0.9966
22.0	0.9978	27.0	0.9965
22.5	0.9977	27.5	0.9964
23.0	0.9975	28.0	0.9962
23.5	0.9974	28.5	0.9961
24.0	0.9973	29.0	0.9959
24.5	0.9972	29.5	0.9958

PROCEDURE

I. Volumetric Pipet

Fill a clean volumetric pipet with distilled water and deliver its contents into a previously weighed container of adequate size. Use Table 2-2 as a guide in deciding on container size and accuracy of weighings. Weigh the container and the dispensed water. Refill the pipet and deliver its contents into the same container. Reweigh and then repeat with a third aliquot of water. Measure the temperature of the water.

TABLE 2-2
Calibration Guide

Pipet Size (ml)	Container Size (ml)	Weighings to (g)
1, 2, 4, 5	25 or 50	±0.001
10	50	±0.01
25	125	±0.01
50	200 or 250	±0.01

II. Buret

Fill a clean buret with distilled water and deliver the excess water into a beaker until the buret reads exactly 0.00. Before you make this adjustment, make certain that there is no air trapped in the buret tip or just above the stopcock. Weigh a clean dry container of adequate size (small Erlenmeyer flask or beaker). For a 50 ml buret, deliver water from the buret into the container until the meniscus reaches the 20.00 ml mark. Reweigh the container. Repeat at 5 ml intervals thereafter (25.00, 30.00, etc) up to 50.00. For a 25 ml buret, the first aliquot should be to the 10.00 ml mark with 2 ml intervals thereafter (12.00, 14.00, etc.) up to 24.00 with 25.00 being the last. All weighings should be to ±0.01 g. Measure the temperature of the water.

CALCULATIONS

In doing your calculations, use Table 2-1 to determine the density of water at your experimental temperature. For buoyancy correction, use 0.001 times the volume to find the correction in grams; this approximation will be accurate to ±0.01 g.

If your results in Part I are not precise to 2 parts per thousand or less, you need practice in the use of the pipet (or practice in weighing and handling of samples to be weighed).

EXERCISES

1. Table 5 on p. 32 gives a corrected weight of water at 27.8°C as 9.98 g. What is the volume occupied by this weight of water?
2. If a buret between 0.00 and 15.00 delivers 14.90 ml of water, what is the correction to be applied?
3. If the correction on a 25 ml pipet is −0.05 ml, what is the true volume delivered by that pipet?
4. The procedure for this experiment suggests that it is not necessary to calibrate a 50 ml buret between 0 and 20 ml. Why not?

(Answers are on p. 242)

REPORT CALIBRATION OF VOLUMETRIC GLASSWARE

I. Volumetric Pipet

Total weight (g)

Empty flask _____

Flask + aliquot 1 _____

Flask + aliquots 1, 2 _____

Flask + aliquots 1, 2, 3 _____

	Weight of aliquot (g)	Buoyancy correction (g)	Corrected weight (g)
Aliquot 1	_____	_____	_____
Aliquot 2	_____	_____	_____
Aliquot 3	_____	_____	_____
		3)	_____
		Average:	_____

1. Water temperature _____ °C
2. Calculate the true volume of water delivered by the pipet using the average corrected weight.

3. Calculate the correction to be applied to the pipet.

II. *Buret*

Buret reading (ml)	Weight of flask + water (g)	Weight of H_2O (g)	Buoyancy correction (g)	Corrected weight (g)	Volume (ml)	Correction (ml)
0.00	_____	——	——	——	0.00	0.00
_____	_____	_____	_____	_____	_____	_____
_____	_____	_____	_____	_____	_____	_____
_____	_____	_____	_____	_____	_____	_____
_____	_____	_____	_____	_____	_____	_____
_____	_____	_____	_____	_____	_____	_____
_____	_____	_____	_____	_____	_____	_____
_____	_____	_____	_____	_____	_____	_____
_____	_____	_____	_____	_____	_____	_____

4. Water temperature _____ °C

5. Plot buret reading (abcissa) versus correction (ordinate).

Name _____ Section _____ Grade _____

Experiment 3
Melting Points

Thermometer. The mass manufacturing of thermometers gives us a product whose accuracy varies with the price we are willing to pay. Thermometers are made from tubing with a uniform size bore running its entire length. A temperature is measured (e.g., the ice point) and a uniformly marked scale is then imprinted onto the tubing to match that reference temperature. For more accurate thermometers, more than one reference temperature is used and the distance between them is equally divided into the appropriate number of parts. Errors arise from nonuniformity of the bore in the tubing and from the care (or lack of care) in determining the reference points. Some thermometers are graduated for *partial immersion* up to a special line on the stem near the bulb. As long as the thermometer is immersed approximately to this line, there will be no serious error due to the rest of the mercury in the column being at a different temperature from that of the mercury in the bulb. When thermometers designed for total immersion are only partially immersed, an error is introduced which becomes significant above 100°C. If your thermometer is a total immersion thermometer, your instructor will show you how to make the necessary stem correction.

In this experiment, you will check the accuracy of your thermometer at 0°C and 100°C. Then assuming a uniform bore, draw a correction graph.

Ice point. The freezing point of pure water at 1 atm pressure is 0.0°C. Impurities in the water will change the freezing point (consult your text regarding colligative properties). The effect of pressure on the melting point of ice is known and is small. It is lowered to −0.008°C when the pressure on the ice-water mixture is increased to 2 atm.

Steam point. The boiling point of pure water at 1 atm pressure is 100.0°C. Impurities in the water again change the boiling point but for equal concentrations the effect is not so pronounced. The effect of pressure on the boiling point is a change of 0.37°C for each torr change in pressure.

Melting points. The melting point of a pure substance is one of its characteristic properties. The temperature range over which it undergoes liquification is called the *melting point range* and for a pure substance is generally 0.5 to 1.0°C. An impure substance will have a larger melting point range and will be lower than that of the pure substance. For a mixture that consists of two or more substances, the melting point will be lower than any of the pure components.

Such properties enable us to determine something about the purity of a compound as well as its identity.

The melting point of a solid may be measured by placing a small quantity in a capillary tube and attaching this tube to the end of a thermometer. Then this is heated by placing it in a suitable bath that is heated gradually until the solid melts; at this time the temperature may be read on the thermometer. If the solids used melt under 100°C, it is possible to use a water bath. For higher melting solids, a mineral oil bath or a metal block heated either electrically or with a flame can be used.

PROCEDURE

A. Thermometer Calibration

To the nearest 0.1°C, check the reading of your thermometer corresponding to the ice point. Use a 150 ml beaker, distilled water, and ice. You will find it necessary to stir the ice-water mixture in order to obtain a constant reading. Record your result on the Report sheet.

To the nearest 0.1°C, check the reading of your thermometer corresponding to the steam point. Insert your thermometer into a specially slit rubber stopper that will fit your 125 ml Erlenmeyer flask. Place distilled water (about 50 ml) in the flask and position the bulb of the thermometer so that it is just above the surface of the water. Bring the water to a gentle boil over a burner and record the results.

B. Melting Points of Solids

Place a small amount of naphthalene in a capillary melting point tube. Do this by pressing the open end of the tube down into the material so that a small amount goes inside the tube. Then turn the tube over and, with the open end up, bang the closed end on the desk top. (Ask your instructor for assistance, if necessary). Only a small amount is needed — about the same volume as that of the glass bead on the closed end of the tube. With your thermometer still in the rubber stopper that you used in A, mount it on a ring stand with your utility clamp. Set up a bath to contain a suitable liquid for heating the samples to be

TABLE 3-1

Melting Points of Selected Compounds

Compound	Melting Point (°C)
Menthol	42
Lauric acid	43
Benzophenone	48
Ethyl carbamate	49
p-Dichlorobenzene	53
Phenyl benzoate	68
Biphenyl	70
Crotonic acid	72
Naphthalene	80
Acenaphthene	95
Citric acid monohydrate	100
Oxalic acid dihydrate	101
Acetanilide	114
dl-Mandelic acid	118
Benzoic acid	121
Succinimide	125
Urea	132
trans-Cinnamic acid	132
Adipic acid	152
Citric acid (anhydrous)	153

measured. If the compounds to be used all melt under 100°C, set up a water bath as described here. If compounds that melt higher than 100°C are to be included, your instructor will furnish you with special instructions. For your water bath, use a 150 ml beaker (It is necessary to use distilled water? Why not?) Attach the capillary tube to the thermometer with an elastic ring so that the solid is right next to the bulb. Immerse the bulb and capillary tube (closed end only of course) in the water. Heat the water with a burner, rapidly at first but more slowly as the melting point is approached. The rate of heating should be no greater than 2°C per minute near the melting point in order to obtain an accurate reading. Record the melting point range — record the temperature at which the solid first begins to melt and the temperature at which melting is complete.

Using the same procedure, determine the melting point range for a mixture of naphthalene and biphenyl.

Obtain an unknown sample from your instructor. It will be one of the substances listed in Table 3-1. Record the unknown number and determine its melting point range. From the melting point range, decide on which of the possibilities that you suspect to be correct. Confirm the identity of your unknown by running a mixed melting point as follows. On a clean watch glass, mix some of the unknown with an equal amount of the compound that you suspect it to be (you may obtain this from your instructor). Determine the melting point range. From your experience in the first part of this section regarding melting points of pure substances and mixtures, decide whether or not you have indeed identified your unknown. If not, repeat the procedure with a different mixture.

EXERCISES

1. In determining the ice point, a thermometer reading of −0.2°C was observed. What is the thermometer error? What is the correction to be applied?
2. The barometric pressure one day was 735.0 torr. What was the boiling point of pure water on that day?
3. If the temperature of the water bath is raised rapidly, the measured melting point will be (A) too high, (B) too low, (C) just right.
4. A student whose thermometer has a correction of −0.5°C in the vicinity of 70°C observed a melting point range of 70.5 to 71.4°C for his unknown. What is his unknown? (A) phenyl benzoate, (B) biphenyl, (C) crotonic acid, (D) not enough information is given.

(Answers are on p.242)

REPORT DETERMINATION OF MELTING POINTS

1. Barometric pressure   752 mm Hg
2. Measured ice point   0.1
3. True ice point   0.0°C
4. Thermometer correction at ice point (No. 3 − No. 2)   −.1°C
5. Measured steam point   99.4°C
6. True steam point at _____ torr   100.0°C
7. Thermometer correction at steam point (No. 6 − No. 5)   −.6°C
8. Make a plot of temperature correction at 0°C and at the steam point (ordinate) versus temperature (abscissa). Draw a straight line between these points. (Refer to p. 6 for help in preparing graphs).
9. Observed melting point range of naphthalene   79.8
10. Corrected range (using graph in No. 8)   79.5
11. Observed melting point range of naphthalene-biphenyl mixture   49.7
12. Observed melting point range of unknown   133
13. Corrected range   132.3
14. Unknown number   132
15. Unknown may be   UREA OR TRANS-CINAMMIC ACID
16. Observed melting point range of unknown mixed with _UREA_   133°
17. Corrected range   132.3

If your results in No. 15 indicate that you have correctly identified your unknown, go directly to No. 21; otherwise try another mixed melting point.

18. Unknown may be   UREA
19. Observed melting point range of unknown mixed with _____   128°−133
20. Corrected range   128.3 − 132.3
21. Unknown is   UREA

Name _____ Section _____ Grade _____

Experiment 4

Formula of a Hydrate

One of the compounds listed in Table 4-1 will be supplied to you. Your task is to identify which one it is by qualitative tests and to determine the number of waters of hydration by quantitative measurements.

TABLE 4-1
Hydrated Salts

$Na_2CO_3 \cdot xH_2O$	$Na_2MoO_4 \cdot xH_2O$	$CuCl_2 \cdot xH_2O$
$CaSO_4 \cdot xH_2O$	$NiSO_4 \cdot xH_2O$	$Na_2HPO_4 \cdot xH_2O$
$BaCl_2 \cdot xH_2O$	$NiCl_2 \cdot xH_2O$	$MnSO_4 \cdot xH_2O$
$CuSO_4 \cdot xH_2O$	$(NH_4)_2C_2O_4 \cdot xH_2O$	$CoCl_2 \cdot xH_2O$

Qualitative Tests. Compounds may be identified as to their component ions by carrying out chemical reactions that give characteristic results; i.e., a precipitate (which may be white or colored) being formed or a gas being evolved. These reactions are usually done in aqueous solution since the component ions will be present in the solution and the reactions of these aqueous ions are often independent of the counter ion present. Some of the characteristic properties and reactions of the ions making up the compounds listed in Table 4-1 will be described.

1. *Solubility.* All the compounds of Table 4-1 are referred to as soluble compounds (at least 2.5 g per 100 ml of solution) except calcium sulfate whose solubility is about 0.3 g per 100 ml of solution. The solubility of calcium sulfate is sufficient to be able to carry out the reactions described since the products (where precipitation occurs) have a much lower solubility.
2. *Color.* Most ions are colorless except for those of the transition elements. A transition metal ion does not always have the same color since the color is dependent on the oxidation number, the coordination number, and often what the ligands are. The characteristic colors in aqueous solution of the ions involved in this experiment are listed in Table 4-2. Occasionally a solid compound will be colored even though its component ions are colorless (e.g., Ag_2CO_3 is yellow). In such instances, the color is attributed to an appreciable amount of covalent character in the bond. Thus the color of solids,

TABLE 4-2
Qualitative Tests

Reagent → Ion ↓	Color	Na_2S	NaOH	$AgNO_3$	$BaCl_2$	$3M\ H_2SO_4$	$18M\ H_2SO_4$	$KMnO_4$
Na^+	Colorless	NR	NR	NR	NR	NR	NR	NR
NH_4^+	Colorless	NR	$NH_3 \uparrow$	NR	NR	NR	NR	NR
Ca^{2+}	Colorless	NR	$Ca(OH)_2 \rightarrow$ white	NR	NR	—	—	NR
Ba^{2+}	Colorless	NR	$Ba(OH)_2 \rightarrow$ white	NR	NR	$BaSO_4 \rightarrow$ white	$BaSO_4 \rightarrow$ white	NR
Cu^{2+}	Blue	$CuS \rightarrow$ black	$Cu(OH)_2 \rightarrow$ blue	NR	NR	NR	NR	NR
Ni^{2+}	Green	$NiS \rightarrow$ black	$Ni(OH)_2 \rightarrow$ green	NR	NR	NR	$NiSO_4 \rightarrow$ yellow	NR
Mn^{2+}	Pale pink	$MnS \rightarrow$ pink	$Mn(OH)_2 \rightarrow$ pink turns brown	NR	NR	NR	NR	$MnO_2 \rightarrow$ brown
Co^{2+}	Pink	$CoS \rightarrow$ black	$Co(OH)_2 \rightarrow$ blue	NR	NR	NR	NR	NR
MoO_4^{2-}	Colorless	$MoS_3 \rightarrow$ brown	NR	$Ag_2MoO_4 \rightarrow$ white sol in HNO_3	$BaMoO_4 \rightarrow$ white sol in HNO_3	NR	$MoO_3 \rightarrow$ yellow	NR
CO_3^{2-}	Colorless	NR	NR	$Ag_2CO_3 \rightarrow$ yellow sol in HNO_3	$BaCO_3 \rightarrow$ white sol in HNO_3	$CO_2 \uparrow$	$CO_2 \uparrow$	NR
Cl^-	Colorless	NR	NR	$AgCl \rightarrow$ white	NR	NR	$HCl \uparrow$	NR(?)
SO_4^{2-}	Colorless	NR	NR	$Ag_2SO_4 \downarrow(?)$ white	$BaSO_4 \rightarrow$ white	NR	NR	NR
$C_2O_4^{2-}$	Colorless	NR	NR	$Ag_2C_2O_4 \rightarrow$ white sol in HNO_3	$BaC_2O_4 \rightarrow$ white sol in HNO_3	NR	RT – NR Δ – $CO + CO_2 \uparrow$	decolorizes RT – slow Δ – faster
PO_4^{3-}	Colorless	NR	NR	$Ag_3PO_4 \rightarrow$ yellow sol in HNO_3	$BaHPO_4 \rightarrow$ white sol in HNO_3	NR	NR	NR

NR means no visible reaction

solutions, and precipitates formed from solution are useful data in establishing the identity of materials.

3. *Reaction with sulfide ion.* In aqueous solution, Na^+, NH_4^+, Ca^{2+}, and Ba^{2+} do not form precipitates with S^{2-} since Na_2S, $(NH_4)_2S$, CaS, and BaS are all soluble in water. Some insoluble metal sulfides that do precipitate are listed in Table 4-2. In the case of MoS_3, occasionally there is a blue color in the supernatant liquid that is some "molybdenum blue" formed by reduction of some of the molybdenum(VI) by the sulfide ion.

4. *Reaction with hydroxide ion.* The source of hydroxide ion for this test is sodium hydroxide. Aqueous ammonia (ammonium hydroxide) is not used since it is also the source of ammonia, a good complexing agent, and thus the reactions are not always the same as with sodium hydroxide. For example, aqueous ammonia reacts with copper(II) ions to form the $[Cu(NH_3)_4]^{2+}$ ion, which has an intense blue color. In contrast, NaOH with copper(II) forms the insoluble blue $Cu(OH)_2$. The metal ions that form insoluble hydroxides are listed in Table 4-2. The ammonium ion (a weak acid) also reacts with hydroxide ion (a strong base) liberating NH_3 gas, which can be detected with moist litmus paper and by its characteristic odor.

5. *Reaction with silver ion.* All the anions used in this experiment form insoluble silver salts as listed in Table 4-2. Silver sulfate is a borderline case and its precipitation only occurs when the ion concentrations are large. Thus you will probably not get a reaction in your tests. Note that two of the silver salts are colored and that two of them are not soluble in 3 M nitric acid. The salts that are soluble in nitric acid are salts of weak acids.

6. *Reaction with barium ion.* All the anions used in this experiment (except Cl^-) form insoluble barium salts, which are listed in Table 4-2. As with the silver salts, the barium salts of weak acids are soluble in nitric acid.

7. *Reaction with dilute sulfuric acid.* There are only two of the ions we are considering that react with 3 M sulfuric acid: Ba^{2+}, which forms insoluble $BaSO_4$, and CO_3^{2-}, which liberates CO_2 gas. What would you expect to happen when silver carbonate or barium carbonate dissolves in dilute nitric acid?

8. *Reaction with concentrated sulfuric acid.* In addition to the same two reactions that occur with dilute sulfuric acid, there are three others with concentrated sulfuric acid as listed in Table 4-2. Note that the oxalate ion does not react at room temperature (RT) but requires some heating.

9. *Reaction with permanganate ion.* The permanganate ion is a good oxidizing agent. In its reactions it is converted either to MnO_2 (brown, insoluble) or to Mn^{2+} depending on the acidity of the solution. In either case, the purple color of the permanganate ion fades as it reacts. The only ions of those under consideration that can be oxidized by permanganate ions are Mn^{2+}, $C_2O_4^{2-}$, and Cl^-. The reaction with oxalate ion is slow at room temperature but is more rapid at higher temperatures. Chloride ion can be oxidized to chlorine but a high concentration of chloride ion is required and you probably will not observe any reaction even with heating (except possible decomposition of the permanganate ion if you "heat it to death").

Quantitative Analysis. The water of hydration may be removed from many hydrated compounds by heating under controlled conditions. Some hydrates decompose when heated. For example, some hydrated metal halides evolve HCl (rather than water), leaving the metal oxide as a product rather than the anhydrous metal halide. Other compounds decompose without reaction with the hydrated water such as ammonium salts, which evolve NH_3 when heated to a high enough temperature.

If one starts with a known weight of a hydrate, carefully dehydrates the compound, and determines the weight of the anhydrous material, it is possible to determine the number of waters of hydration. This necessitates knowing the formula of the anhydrous salt and of course the formula of water. A calculation of this type is included in the exercises and is the same sort of determination you are to do on your hitherto unknown material.

PROCEDURE

A. Qualitative Tests

Carry out the necessary chemical tests to identify your compound. You should do this during time periods in Part B when you are waiting for materials to heat up or cool down. Prepare a solution of your unknown by dissolving about 0.2 g of the solid in 20 ml of water. It will be necessary to stir the water to dissolve the sample. If it doesn't all dissolve, allow the excess solid to settle and decant the solution away from the solid. Each test will require 1 ml (20 drops) of this solution except two tests done directly on the solid.

For heating solutions, set up a water bath using a 250 ml beaker and your aluminum bath rack. Fill the beaker with water and heat it over a flame. This heats the solution gently and there is no danger of boiling over. When a solution in a test tube is heated directly over a flame, there is a great tendency for it to "bump."

Whenever you add one solution to another, they will not mix completely by themselves (unless you are willing to wait for several months). You must be the mixer by shaking or stirring with a stirring rod. Record your observations on the Report sheet.

1. *Solubility.* From your observations in preparing a solution as described above, would you classify your sample as soluble or not so soluble?
2. *Color.* What color is your solid sample? What color solution does it form?
3. *Reaction with sulfide ion.* To 1 ml of a solution of your unknown, add 3 drops of 2 M Na_2S. If a precipitate forms, you may want to centrifuge it to observe its color better.
4. *Reaction with hydroxide ion.* To 1 ml of a solution of your unknown, add 3 drops of 2 M NaOH. If no precipitate forms, you should check for ammonia being evolved. Do this by heating the solution of your unknown to which NaOH has been added in a water bath and holding a moistened (with distilled water) piece of red or neutral litmus paper over the mouth of the test tube. If the vapors escaping from the test tube turn the litmus blue, that is a positive test for the ammonium ion.
5. *Reaction with silver ion.* To 1 ml of a solution of your unknown, add 3 to 5 drops of 0.1 M $AgNO_3$. If a precipitate forms, centrifuge, decant (discard the decantate), and add 10 drops of 3 M HNO_3 to the solid. Stir to check solubility.
6. *Reaction with barium ion.* To 1 ml of a solution of your unknown, add 3 to 5 drops of 0.5 M $BaCl_2$. If a precipitate forms, check its solubility in 3 M HNO_3 as in No. 5.
7. *Reaction with dilute sulfuric acid.* Place 5 to 10 mg of your solid unknown in a test tube. Add 5 drops of 3 M H_2SO_4 directly to the solid.
8. *Reaction with concentrated sulfuric acid.* CAUTION: Sulfuric acid is especially dangerous; it causes burns when it comes in contact with skin. Take special care to protect your eyes and don't point your test tube at anyone (including yourself). Place 5 to 10 mg of your solid unknown in a test tube. Add 5 drops of 18 M H_2SO_4 directly to the solid. If there is no apparent reaction, CAREFULLY warm over a small flame.
9. *Reaction with permanganate ion.* Add 1 drop of 3 M H_2SO_4 and 2 drops of 0.01 M $KMnO_4$ to 1 ml of a solution of your unknown. If no apparent reaction occurs, heat in a water bath.

B. Quantitative Analysis

Your instructor will tell you whether to heat your sample in an oven or over a flame. Follow the appropriate procedure below. The directions here are for a single analysis; do the analysis in duplicate if so instructed.

Oven drying. Clean your crucible, place it in a beaker, and put it all in a 120°C oven. (How are you going to tell which is yours?) After at least 30 minutes, remove it (How are you going to handle the hot beaker?) and place the crucible in a desiccator. When it has cooled to room temperature (about 20 minutes), weigh to the nearest 0.1 mg. See pp. 19, 23, and 25 for further instructions. Place between 1 and 2 g of your unknown in the crucible and reweigh. Heat the sample in the oven for at least 1 hour. Cool in a desiccator and weigh. Calculate percent water in the sample and the number of water molecules per formula unit.

Flame drying. Clean your crucible, place it on a triangle, and heat strongly with a flame for 5 minutes. Place the crucible in a desiccator to cool. Leave the desiccator cover ajar for 2 or 3 minutes allowing the contents to cool somewhat before putting the cover on tightly. When it has cooled to room temperature (about 30 minutes), weigh to the nearest 0.1 mg. See pp. 19, 23, and 25 for further instructions. Place between 1 and 2 g of your unknown in the crucible and reweigh. Heat the sample over a flame, gently at first since as the water evaporates, it might cause some spattering. If there is an obvious change such as in color or in opaqueness, continue heating gently until the change is complete. If there is no such visible change, heat gently for 5 minutes and then strongly for 5 minutes. Cool in a desiccator as before and weigh. Calculate the percent water in the sample and the number of water molecules per formula unit.

EXERCISES

1. A white compound composed of ions listed in Table 4-2 is soluble in water forming a colorless solution. With barium chloride, a white precipitate forms that is insoluble in nitric acid. There is no reaction with $AgNO_3$, H_2S, or NaOH. What is the compound?
2. A pale pink compound composed of ions listed in Table 4-2 is soluble in water forming what appears to be a colorless solution. With silver nitrate, a white precipitate forms that is insoluble in nitric acid. A gas is evolved with concentrated sulfuric acid but not with dilute sulfuric acid. With H_2S, a pink precipitate forms. What is the compound?
3. A sample weighing 2.0526 g loses 0.1534 g upon heating. What is the percent weight loss?
4. A substance is identified as a hydrate of magnesium sulfate. A sample is weighed out, heated, and reweighed. Using the data given, calculate the percent of water in the substance and the number of water molecules per formula unit.

	Before drying	*After drying*
Weight container + sample	20.5046 g	19.4983 g
Weight container	18.5421 g	

(Answers are on p. 242)

REPORT FORMULA OF A HYDRATE

1. Unknown number _____
2. Qualitative Tests

Test	Observation	Conclusion
1. Solubility		
2. Color	Solid: Solution:	
3. H_2S		
4. NaOH		
5. $AgNO_3$	 HNO_3 solubility:	
6. $BaCl_2$	 HNO_3 solubility:	
7. 3 M H_2SO_4		
8. 18 M H_2SO_4		
9. $KMnO_4$		

3. Unknown is identified as _____
4. Write balanced equations for all reactions occurring in the tests described in No. 2.

	Sample 1	Sample 2
5. Weight of crucible, undried sample (g)	_____	_____
6. Weight of crucible (g)	_____	_____
7. Weight of undried sample (g)	_____	_____
8. Weight of crucible, dried sample (g)	_____	_____
9. Weight of crucible (g) (No. 6)	_____	_____
10. Weight of dried sample (g)	_____	_____
11. Weight of water lost (g)	_____	_____

12. Calculate the percent of water in the sample:

$$\frac{\text{water lost}}{\text{wt. dried sample}} \times 100 = \%$$

13. Calculate the number of water molecules per formula unit of the hydrate:

$$\frac{\text{water lost gms}}{\text{mole wt } H_2O} = \text{moles water lost}$$

$$\frac{\text{weight dried sample}}{\text{molecular wt of dried sample}} = \text{moles unknown}$$

$$\frac{\text{moles water lost}}{\text{moles unknown}}$$

14. During the dehydration of the unknown sample, loss of sample from the crucible due to spattering could cause a serious error.

   a. Can the magnitude (size) of this error be estimated? Explain.

   b. If a significant amount of sample is lost, will the determined percentage of water in the sample be too high or too low?

Name _____ Section _____ Grade _____

Experiment 5
The Chemistry of Nitric Acid

Nitric acid reacts with most of the elements but the conditions necessary for reaction and the products formed vary. This experiment is concerned with the reactions of metals with nitric acid and the thermal decomposition of the products formed.

Most metals react with nitric acid easily to form the corresponding metal nitrates. When a metal can form more than one oxidation state, the product is governed by the concentration of the nitric acid — the more concentrated acid giving the higher oxidation state. Some metals (or rather metalloids) do not form stable nitrates in solution and thus the product obtained is a hydrated oxide. A few metals dissolve well in dilute nitric acid but become passive when concentrated nitric acid is used. This passiveness is attributed to the formation of a coating (probably an oxide) that prevents further reaction. The reactions of a variety of metals are summarized in Table 5-1.

TABLE 5-1

Reactions of Metals with Nitric Acid

Metal	Oxidation Product
Li, Na, K	Li^+, Na^+, K^+
Mg, Ca	Mg^{2+}, Ca^{2+}
Mn, Co, Ni, Cu	Mn^{2+}, Co^{2+}, Ni^{2+}, Cu^{2+}
Fe, Al, Cr	Passive to concentrated acid
Fe	Fe^{2+} or Fe^{3+} (depends on concentration)
Al, Cr	Al^{3+}, Cr^{3+} (with dilute acid)
Zn, Cd	Zn^{2+}, Cd^{2+}
Sn, Sb	$SnO_2 \cdot xH_2O$, $Sb_2O_5 \cdot xH_2O$
Pb, Bi	Pb^{2+}, Bi^{3+}

The reduction products of the nitric acid vary with the activity of the metal and the concentration of the acid. Concentrated nitric acid gives NO_2 as the principal product. As the concentration of the acid decreases, the extent of reduction of the nitrogen increases with typical products including NO, N_2, N_2H_4, and NH_3. The more active the metal, the greater the reduction tends to be and metals well above hydrogen in the activity series may also evolve some H_2 with dilute acid. Some typical reactions are given here but in all cases the reaction shown is the principle reaction but not necessarily the only reaction occurring.

Copper + concentrated nitric acid:
$$Cu + 4\,HNO_3 \longrightarrow Cu(NO_3)_2 + 2\,NO_2 + 2\,H_2O$$

Copper + dilute (6 M) nitric acid:
$$3\,Cu + 8\,HNO_3 \longrightarrow 3\,Cu(NO_3)_2 + 2\,NO + 4\,H_2O$$

Magnesium + dilute (3 M) nitric acid:
$$4\,Mg + 10\,HNO_3 \longrightarrow 4\,Mg(NO_3)_2 + NH_4NO_3 + 3\,H_2O$$

Magnesium + dilute (1%) nitric acid:
$$Mg + 2\,HNO_3 \longrightarrow Mg(NO_3)_2 + H_2$$

Metal nitrates decompose upon heating but not all give the same products. These reactions are summarized in Table 5-2. The factors governing the product formed are the relative stabilities of the metal nitrite or the metal oxide at the temperature of decomposition. Exercise 5 illustrates the effect of temperature on the stability of silver oxide.

TABLE 5-2

Thermal Decomposition of Metal Nitrates

Metal Nitrate	Products
IA and IIA metals	Metal nitrite + O_2
Others except noble metals	Metal oxide + NO_2 + O_2
Noble metals	Metal + NO_2 + O_2

In this experiment you will be given a sample of a metal. The metal is to be reacted with nitric acid and the resulting solution (or mixture if the metal is tin or antimony) evaporated to dryness. From your observations during the reactions and from the weight of product formed, you are to identify the sample. The metal supplied to you will be one of those listed in Table 5-1 with a few exceptions. Metals such as lithium, sodium, and potassium react too vigorously, or even violently, to be suitable for use in this experiment. Other metals that react too slowly to allow completion of the experiment within the allotted time will not be used.

TABLE 5-3

Chemistry of Metal Nitrates

Metal Nitrate	Color of HNO_3 Solution	Final Product	Color of Final Product
$LiNO_3$, $NaNO_3$, KNO_3	Colorless	$LiNO_2$, $NaNO_2$, KNO_2	White
$Mg(NO_3)_2$, $Ca(NO_3)_2$	Colorless	$Mg(NO_2)_2$, $Ca(NO_2)_2$	White
$Mn(NO_3)_2$	Pale pink	Mn_2O_3	Black
$Co(NO_3)_2$	Violet	Co_3O_4	Black
$Ni(NO_3)_2$	Green	NiO	Black
$Cu(NO_3)_2$	Blue	CuO	Black
$Fe(NO_3)_3$	Yellow	Fe_2O_3	Red-brown
$Al(NO_3)_3$	Colorless	Al_2O_3	White
$Cr(NO_3)_3$	Blue-violet	Cr_2O_3	Green
$Zn(NO_3)_2$	Colorless	ZnO	White
$Cd(NO_3)_2$	Colorless	CdO	Brown
$SnO_2 \cdot xH_2O$	White solid	SnO_2	White
$Sb_2O_5 \cdot xH_2O$	White solid	Sb_2O_5	White
$Pb(NO_3)_2$	Colorless	PbO	Yellow
$Bi(NO_3)_3$	Colorless	Bi_2O_3	Gray-black

PROCEDURE

CAUTION: Nitric acid is dangerous! Be careful, wear your safety glasses, and be very cautious.

A. Cleaning and Drying the Crucible

Place your triangle on an iron ring clamped to a ring stand. Add 1 or 2 ml of concentrated nitric acid to a crucible and warm it gently in order to clean the crucible. When it is cool enough to handle, rinse well with tap water followed by distilled water, and then dry. Support the crucible, with cover ajar, in the triangle and heat it strongly for at least 5 minutes. Allow to cool in a desiccator leaving the cover ajar slightly until it has cooled somewhat. It will take about 15 minutes for the crucible to cool.

B. Preliminary Tests

While the crucible is heating and then cooling, perform the following tests. These tests need to be done in order to help you decide on the best method for dissolving your sample (both for ease of dissolving and for safety) and to help you identify the metal.

Place a very small amount of your sample (about 25 mg) in a 30 ml test tube. Add about 5 ml of water. If there is no reaction, warm the test tube gently (do not heat strongly; it might "blast off" or, as a chemist would say, "bump"). If there is still no reaction or if the reaction is very mild, discard and repeat with $3\,M\,HNO_3$. If there is still no reaction or if the reaction is very mild, discard and repeat with $6\,M\,HNO_3$. If necessary, repeat again with concentrated HNO_3; however, there should be no need to try heat.

Record your observations regarding reaction conditions and any other pertinent facts such as color of the solution formed and of gases evolved.

C. Weighing and Treatment of Sample

When the crucible is cool, weigh it and its cover to the nearest 0.1 mg. Add about 0.5 g of sample to the crucible and weigh.

Support the crucible and contents on the triangle and carefully add dropwise either water or the proper concentration of nitric acid as determined in B. After adding each drop, let the reaction subside before adding more. It may be necessary to heat gently or to change to more concentrated nitric acid as the volume of acid in the crucible increases.

When the metal sample is completely dissolved, place the crucible cover ajar and begin heating *gently* with a low flame. Stop heating immediately if you observe evidence of boiling over or spattering; allow to cool slightly and then continue gentle heating. (*Note:* Spattering causes errors due to sample loss.) Continue heating in this manner until all the liquid has evaporated and the residue in the crucible is dry. Now heat strongly for about 5 minutes. Cool as before and weigh.

EXERCISES

1. What volume of nitric acid would be theoretically needed to react with 0.50 g of magnesium (a) if concentrated HNO_3 is used and NO_2 is the only reduction product? (b) if $3\,M$ HNO_3 is used and NO is the only reduction product?
2. What volume of concentrated nitric acid would be theoretically needed to react with 0.50 g of lead if NO_2 is the only reduction product?
3. What weight of TiO_2 could theoretically be formed from 0.5000 g of titanium?
4. If you have 5 ml of a solution and you lose 1 drop of the solution, what percentage of the solution has been lost? (Assume 20 drops per ml).

5. For the reaction $Ag_2O\ (s) \longrightarrow 2\ Ag\ (s) + \frac{1}{2} O_2\ (g)$, at 25°C and 1 atm pressure, $\Delta H = +7.31$ kcal and $\Delta S = +0.0158$ kcal/deg. Assuming that ΔH and ΔS are independent of temperature, determine (a) whether or not the reaction is spontaneous at 25°C and at 300°C and (b) the temperature at which the system is at equilibrium.

(Answers are on p. 243)

REPORT CHEMISTRY OF NITRIC ACID

1. Record your observations during the preliminary tests.

   _____

   _____

   _____

   _____

2. Unknown number _____

3. Weight of crucible, cover, and sample (g)   _____

   Weight of empty crucible and cover (g)   _____

   Weight of sample (g)   _____

   Weight of crucible, cover, and product (g)   _____

   Weight of empty crucible and cover (from above) (g) _____

   Weight of product (g)   _____

4. The color of the final product is   _____

5. The metal sample is suspected to be   _____

   The final product is thus suspected to be   _____

6. Write balanced equations for

   (a) the reaction of the metal with nitric acid

   (b) the thermal decomposition of the product formed in 6(a):

7. Calculate the theoretical weight of final product that could be formed from your sample if it is the metal that you suspect it to be.

8. If the calculation in No. 7 does not confirm your answer to No. 5, what would you now suspect the metal sample to be? _____

   The final product from that metal would be _____

   If the calculation in No. 7 confirms your answer to No. 5, omit No. 8, 9, and 10.

9. Write balanced equations for

   (a) the reaction of the metal with nitric acid:

   (b) the thermal decomposition of the product formed in 9(a):

10. Calculate the theoretical weight of final product that could be formed from your sample based on the reactions in No. 9.

11. Calculate the percent error in the amount of final product actually formed and recovered.

Name _____ Section _____ Grade _____

Experiment 6

Paper Chromatography

Substances commonly occur in mixtures and often it is desirable to separate them into their pure components. A substance desired may be mixed with large amounts of one or more other substances or it may contain only a small amount of impurity. The traditional methods of separating such mixtures are based on differences in vapor pressures (separation can then be achieved by distillation or sublimation) or on differences in solubilities (separation can then be achieved by crystallization). The technique of chromatography is able to effect such separations better and faster than these traditional methods. Chromatography can unmix a complex mixture in minutes and it is applicable to a wide variety of substances such as gases, metal ions, isomers, simple organic molecules, and complicated biomolecules. It is this versatility and speed that has made chromatography such a powerful tool for the scientist.

The mechanics of chromatography are rather straightforward. The mixture to be separated is introduced into a flowing fluid (a gas or a liquid) that is passing through a stationary sorptive material. The sorptive material has an attraction for the substances in the mixture and the substances migrate along the sorptive material in the direction of flow at a rate governed by the affinity for the sorptive material. There are two opposing forces acting on the substances, a driving force and an impeding force. The driving force is determined by the relative solubilities in the flowing solvent; the more soluble the substance, the more easily it moves along the sorptive material. The impeding action is determined by the relative attraction of the sorptive material for the substances to be separated; the greater the attractive forces, the more slowly the substance moves along the sorptive material.

The actual means by which a chromatographic separation is carried out is dictated by the nature of the substances to be separated and the nature of the sorptive material to be used. This experiment utilizes paper chromatography; Experiment 7 uses thin-layer chromatography. Ion-exchange chromatography, a type of column chromatography, is used in Experiment 8. (Another widely used type is gas chromatography). In all these experiments, the separations are effected for the purpose of identification and are thus referred to as an *analytical* use of chromatography. If the separations are carried out for the purpose of purification, it is then called *preparative* chromatography. Even though the theory of chromatography has been well studied, it is not completely understood. Much of the practical use of chromatography is trial and error with the procedures used being modifications of existing procedures.

Cellulose paper is a very complex medium. There are regions that are highly ordered, the polysaccharide chains being held together by hydrogen bonds. The water in these regions of the paper is tightly bonded to the paper and is not able to dissolve solutes very well. Other regions of the paper are less ordered and the water in these regions is more like solvent water and thus able to dissolve solutes. The paper contains OH groups that can bond weakly to solutes through the oxygen or through hydrogen bonds. In addition the surface of the paper tends to have a negative charge that will result in an attractive force for positive species.

In this experiment, a mixture of metal ions will be separated. Known mixtures will be separated and identified by the formation of colored species by chemical reactions and R_f values. Then an unknown mixture will be separated and the components identified by the same chemical tests and by comparing the R_f values with those of the knowns. It would be expected that the metal ions would be soluble in the water medium of the paper and that the hydrated metal ions would also be attracted to the oxygen and hydroxyl group of the paper through hydrogen bonds. This results in a strong attraction of the paper for hydrated metal cations. The solvent to be used is a butyl alcohol-acetone-hydrochloric acid mixture. This phase is much less polar than water and on this basis the solubilities of the metal ions would be much less. If polarity were the only factor, there would not be much separation of the metal ions since their attraction for the paper and its aqueous medium would be so much greater than their solubilities in the solvent. The solubilities of some of the metal ions however, are increased by the presence of the hydrochloric acid. Metal ions that form chloro-complexes will be quite soluble in the solvent and the solubility will be related to the ease of formation of such complexes; the more stable the complex, the greater the solubility.

Although chemists have been able to prepare chloro-complexes of all the metal ions in this experiment, under the conditions of this experiment some do not form chloro-complexes at all and others form them to varying degrees. During the separation of known mixtures you will be able to see the formation of the chloro-complexes of three of the metal ions and we shall use these to illustrate the principles involved.

$$Fe(H_2O)_6^{3+} + 4\,Cl^- \rightleftharpoons FeCl_4^- + 6\,H_2O$$
yellow

$$Co(H_2O)_6^{2+} + 4\,Cl^- \rightleftharpoons CoCl_4^{2-} + 6\,H_2O$$
blue

$$Cu(H_2O)_6^{2+} + 4\,Cl^- \rightleftharpoons CuCl_4^{2-} + 6\,H_2O$$
yellow

The colors of these chloro-complexes are more intense than the colors of the hydrated metal ions and thus are visible. (The hydrated metal ions are colored but are present on the paper in such dilute concentrations that they appear colorless.) Iron(III) forms the most stable of these chloro-complexes and is found to move along with the solvent very close to the leading edge of the solvent as it moves across the paper (called the *solvent front*). Cobalt(II) also forms one and consequently moves well with the solvent but not as far as the iron(III) does. Copper(II) forms a chloro-complex of comparable stability to cobalt(II) and is found in the same region of the chromatogram as cobalt(II). Thus you can see the relationship between chloro-complex formation and relative distance moved in the development of the chromatogram.

When the chromatogram is fumed with ammonia, the hydrochloric acid is neutralized.

$$HCl + NH_3 \longrightarrow NH_4Cl$$

In addition, some color changes will be observed due to reactions of the metal ions with the ammonia.

$$CuCl_4^{2-} + 4\,NH_3 + 2\,H_2O \longrightarrow Cu(NH_3)_4(H_2O)_2^{2+} + 4\,Cl^-$$
deep blue

$$FeCl_4^- + 3\,NH_3 + 3\,H_2O \longrightarrow \underset{\text{yellow-brown}}{Fe(OH)_3} + 3\,NH_4^+ + 4\,Cl^-$$

$$CoCl_4^{2-} + 6\,NH_3 \longrightarrow Co(NH_3)_6^{2+} + 4\,Cl^-$$

$$4\,Co(NH_3)_6^{2+} + \underset{\text{from air}}{O_2} + 2\,H_2O \longrightarrow \underset{\text{yellow}}{Co(NH_3)_6^{3+}} + 4\,OH^-$$

With potassium iodide reagent, the following reactions occur.

$$Pb^{2+} + 2\,I^- \longrightarrow \underset{\text{yellow}}{PbI_2}$$

The copper(II) is a good enough oxidizing agent to oxidize I^- to I_2; it is reduced to copper(I).

$$2\,Cu^{2+} + 4\,I^- \longrightarrow \underset{\text{red-brown}}{2\,CuI + I_2}$$

With sodium sulfide reagent, the following reactions occur.

$$Pb^{2+} + S^{2-} \longrightarrow \underset{\text{black}}{PbS}$$

$$Cd^{2+} + S^{2-} \longrightarrow \underset{\text{yellow}}{CdS}$$

$$Cu^{2+} + S^{2-} \longrightarrow \underset{\text{black}}{CuS}$$

In addition, NiS, CoS, and FeS are all black.

With ammonium thiocyanate in acetone reagent, the visible reactions are as follows.

$$Co^{2+} + 4\,SCN^- \longrightarrow \underset{\text{blue}}{Co(SCN)_4^{2-}}$$

$$Fe^{3+} + SCN^- \longrightarrow \underset{\text{red}}{Fe(SCN)^{2+}}$$

With dimethylglyoxime (DMG) reagent, nickel(II) forms a red compound. The reaction is described on p.102. Co(II) and Fe(III) form similar compounds but the colors are similar (yellow-brown) and are not very useful as confirmatory tests for these ions.

The extent to which a solute is attracted to the sorbent material is expressed by the *retardation factor*, R_f.

$$R_f = \frac{\text{distance solute has moved}}{\text{distance solvent has moved}}$$

It is simply the movement of the substance relative to the movement of the solvent during the same period of time. In a good separation, each component will have its own characteristic R_f value. The R_f value obtained for a given compound, however, is dependent on the solvent system, sorbent material used, temperature, and concentration of the sample.

Therefore chromatograms of unknowns are usually run under identical or nearly identical conditions as knowns. In this experiment the conditions of the development should be nearly identical and R_f values can be used as an aid in identification of the metal ions present in an unknown mixture.

PROCEDURE

Mark two pieces of 11 cm Whatman No. 1 filter paper with a pencil as shown in Fig. 6-1. Cut a slit in the paper using a sharp knife or razor blade to hold a wick prepared as shown in Fig. 6-2(a). Use an ordinary graphite pencil to mark the paper; felt, fiber tip, ball-point, or fountain pens cannot be used. On one piece of filter paper, label two spots *A* and two spots *B* as in Fig. 6-1; on the other piece of filter paper, label all spots *UNK* (unknown).

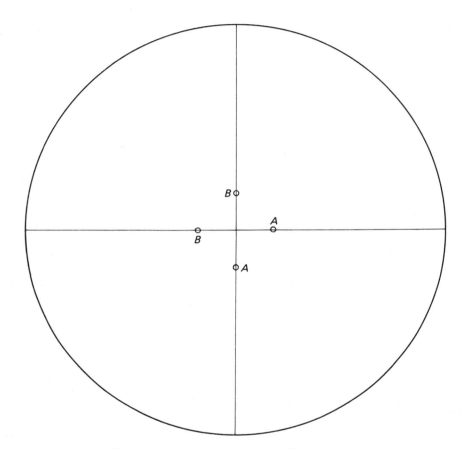

Figure 6-1. Diagram for marking filter paper.

Use a 3 mm diameter glass stirring rod for spotting the paper. Make a spot by dipping the end of the clean stirring rod into the solution and touching it quickly and lightly to the paper. The spot should be 3 to 5 mm in diameter. You may wish to practice with water and a small strip of filter paper before you actually spot the solutions onto the circular filter paper. If you find you are getting too much solution on the end of the stirring rod, try briefly touching the end of the rod to the side of the container before touching it to the paper. After making a spot, let it dry and then apply another drop of the same solution to the spot. Allow this to dry also before proceeding. Be sure to clean the glass rod each time you use a different solution.

Solution A contains $Cu(NO_3)_2$, $Cd(NO_3)_2$, and $Pb(NO_3)_2$ all at $0.10\ M$ concentrations. Solution B contains $0.10\ M\ Fe(NO_3)_3$, $0.10\ M\ Co(NO_3)_2$, and $0.20\ M\ Ni(NO_3)_2$. Your unknown will contain at least one but no more than four of these six ions. There should be two spots of solution A and two spots of solution B on one chromatogram and four spots of the unknown on the other chromatogram.

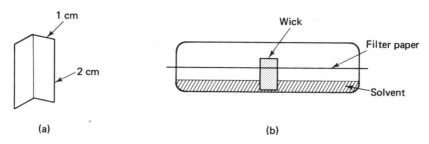

Figure 6-2. (a) Dimensions of wick made from 2 × 2 cm filter paper. (b) Filter paper in place between Petri dish halves.

When the spots have dried, set up the apparatus shown in Fig. 6-2(b), which uses *either* two Petri dish tops (100 × 20 mm) *or* two Petri dish bottoms (95 × 18 mm). The wick should just reach the bottom of the dish. When everything is adjusted properly, place freshly prepared developing solvent (prepared by mixing 10 ml *n*-butanol + 10 ml acetone + 1 ml $12\ M$ HCl) into one Petri dish. Place the filter paper on top of the dish with the wick in the solvent and cover with the other Petri dish. Let this stand until the solvent nearly reaches the edge of the Petri dish. Remove the cover and with a pencil quickly mark the solvent front along each of the four radii (try not to get your fingers onto the wet part of the paper or into the solvent; if you do, you may find that the paint on your pencil is soluble in the solvent). While this chromatogram is drying, develop the second chromatogram using the same solvent (it should *not* be necessary to prepare a new portion of solvent). Notice any colors in the chromatograms as they develop; when the solvent evaporates, the colors may fade.

When the chromatograms are dry, note any colors and mark the locations of those spots with a pencil. Then place the chromatogram in a jar containing a cotton ball moistened with a few drops of $15\ M$ aqueous ammonia. The reactions with NH_3 will take only 1 or 2 minutes; again, make note of any colors in the chromatograms and mark their locations.

There are four reagents to be used for identification tests and they are to be applied along the marked lines (one reagent per line) using droppers. The reagent is applied by moving the tip of the dropper along the line and at the same time allowing the reagent to flow from the dropper. The width of the band should be at least 1 cm with the pencil line as the center. Test one of the chromatograms of solution A with $0.5\ M$ KI and the other with $2\ M\ Na_2S$. Test one of the chromatograms of solution B with 10% NH_4SCN in acetone and the other with $0.1\ M$ dimethylglyoxime in 95% ethanol. Of the four chromatograms of the unknown, test each with a different one of the four reagents. When they are dry, record the colors and mark the location of each colored spot with a pencil. Some of the tests may be slow in developing the characteristic color but all should be evident by the time the paper is dry.

Measure the location of each spot and calculate R_f values for each ion. See p. 96 for instructions on measurement of R_f values.

EXERCISES

1. Why is a pencil used to mark the chromatograms rather than a pen?

2. What is the R_f value of the substance in the chromatogram shown below?

3. From the results of a chromatographic analysis involving the metal ions considered in this experiment, a student thought only one ion was present in his unknown. The results of the tests on the chromatogram were inconclusive so the following tests were run on the original solution directly. With Na_2S, a yellow precipitate formed. There was no reaction on individual portions of the solution with KI, NH_4SCN, or dimethylglyoxime (with added base). What ion or ions were present?

4. Another student in the same predicament as the one in Exercise 3 tried the same approach. With Na_2S, a black precipitate formed. NH_4SCN and dimethylglyoxime gave no visible reaction. What ion or ions were present?

(Answers are on p.244)

REPORT PAPER CHROMATOGRAPHY

1. Unknown number _____
2. Color of unknown solution _____
3. Solution *A*

Ion	Color with				Distance (mm)		R_f
	Solvent	NH_3	KI	Na_2S	Spot	Solvent	
Cu^{2+}							
Cd^{2+}							
Pb^{2+}							

4. Solution *B*

Ion	Color with				Distance (mm)		R_f
	Solvent	NH_3	NH_4SCN	DMG	Spot	Solvent	
Fe^{3+}							
Co^{2+}							
Ni^{2+}							

5. Unknown

Ion	Color with					
	Solvent	NH_3	KI	Na_2S	NH_4SCN	DMG
A						
B						
C						
D						

| | Distance (mm) | | |
Ion	Spot	Solvent	R_f
A			
B			
C			
D			

6. Unknown contains _____

7. On the basis of your results, which forms the more stable chloro anion, cobalt(II) or nickel(II)? How did you arrive at your conclusion?

Name _____ Section _____ Grade _____

Experiment 7

Thin-Layer Chromatography

Thin-layer chromatography is a method for the separation of a mixture into its components on a very thin layer of material. The separation can occur by adsorption, ion-exchange, or partition (between solvent and stationary phase) processes. Thin-layer chromatography (TLC) has come into widespread use because it is one of the simplest, most convenient, and fastest of the chromatographic procedures. It is possible to separate closely related substances using this technique. For example the *cis* and *trans* isomers of a compound can be separated as can two organic acids that differ from each other by only one carbon atom.

Thin-layer chromatography is used as a rapid check of the identity or purity of the products of a reaction. It is used in criminology for the detection of drugs, narcotics, and poisons. In industry it is sometimes used in quality control. It is an excellent technique when only micro samples are available, as is the case many times in biochemical preparations.

An introduction to chromatography and its application using paper (cellulose) is given in the first four paragraphs of Experiment 6. You should read that before continuing on with this experiment.

The separations are carried out on a thin layer of material of uniform thickness, usually 100 to 250 μ thick (1 micron, μ, is 1×10^{-4} cm), which is supported on a glass plate or a plastic sheet. The most commonly used adsorbents are cellulose, silica (SiO_2) gel, and alumina (Al_2O_3). When cellulose is the adsorbent, thin-layer chromatography is similar to paper chromatography in principle as well as in technique, except that TLC gives much better resolution of the components. The high resolution of substances is due to the small diffusion rate in the adsorbent. The adsorbents do not adhere very well to glass plates and frequently a binder, such as calcium sulfate or starch, is added to improve adhesion. The plates are coated by spreading a slurry of adsorbent on them in a thin, uniform layer, usually with a commercial applicator. It is possible to buy plates that have been precoated with the most commonly used adsorbents.

The extent to which a substance has been adsorbed by the adsorbent is expressed by the retardation factor, R_f.

$$R_f = \frac{\text{distance solute has moved}}{\text{distance solvent has moved}} \quad (1)$$

It is simply the movement of the substance relative to the movement of the solvent during the same period of time. In a good separation, each component will have its own characteristic R_f value. The R_f value obtained for a given compound is dependent, however, on the solvent system, the adsorbent used, and such diverse factors as the thickness of the adsorbent, the concentration of the sample, the size of the spot, and the method of development. Therefore unless chromatograms are run under identical or nearly identical conditions, it is impossible to compare the R_f values obtained. For this reason the known substances are frequently run along with the mixture that is being separated. In this experiment the conditions of the development should be nearly identical and you can use the R_f values as a means of identification of the substances present in an unknown mixture. Yet, you may still find that the R_f values for the same substance will differ from each other by 0.01 or 0.02.

You will investigate the separation of a series of aliphatic organic acids that differ from each other by the number of carbon atoms they contain. The organic acids that contain from two to six carbon atoms can be completely separated from each other by thin-layer chromatography on cellulose plates.[1] The names and formulas are listed in Table 7-1. A plastic sheet precoated with cellulose will be used as the medium for the separation. Cellulose is a

TABLE 7-1

Aliphatic Acids

Number of Carbon Atoms	Name	Formula
2	Ethanoic (acetic) acid	CH_3COOH
3	Propanoic (propionic) acid	CH_3CH_2COOH
4	Butanoic (butyric) acid	$CH_3(CH_2)_2COOH$
5	Pentanoic (valeric) acid	$CH_3(CH_2)_3COOH$
6	Hexanoic (caproic) acid	$CH_3(CH_2)_4COOH$

The common name of the acid is given in parentheses.

naturally occurring polymer consisting of long chains of glucose units. It is hydrophilic (liking water) and therefore is used for the separation of polar, water-soluble molecules. In aqueous solvents the cellulose becomes swollen with water molecules, expanding its polymeric network. The carboxylate groups of the acids are attracted to the polar groups of the cellulosic network by dipole-dipole forces and this interaction restricts the movement of the acids through the cellulose. On the other hand the hydrocarbon chain of the acids interacts with the organic component of the solvent, which causes the acids to move along with the solvent.

The separation of the organic acids listed in Table 7-1 is achieved mainly because of (1) the size differences of the molecules and (2) their relative interactions with the organic solvent. The acids containing the larger numbers of carbon atoms are larger in size and cannot diffuse as far into the cellulosic network as the smaller acids. Therefore the larger acids spend a greater proportion of the time in the moving solvent phase and a lesser proportion of the time adsorbed by the cellulose than do the smaller acids. As the number of carbon atoms in the hydrocarbon chain increases, the attractive interaction between the acid molecule and the organic solvent increases, again causing the larger acids to spend a greater proportion of the time in the moving solvent phase than do the smaller acids.

The solvent system used in the development of the chromatogram is a solution of n-propanol, 15 M aqueous ammonia, and water, present in the ratio of 9:1:2 by volumes. The acids are spotted onto the cellulose plate in the form of very dilute solutions (containing about 4 mg of acid per ml of water) of their ammonium salts.* After development and drying of the chromatogram, the acid spots are located by spraying the plate with a solution of bromophenol blue, which is an acid-base indicator. The acidic color of the indicator is yellow and the basic color is blue with the color change occurring in the pH range of 3.0 to 4.6.

Acetic acid is the "active" ingredient of vinegar. You may not find its odor unpleasant but some of the other members of this group of organic acids are vile smelling substances. Although you will be using very dilute solutions, the odor is still strong enough to detect and it would be wise for aesthetic reasons to avoid getting them on your skin or clothing.

*These solutions are prepared by dissolving the acid in water to give the desired concentration and then neutralizing with a twofold excess of aqueous NH_3.

PROCEDURE

Obtain two 6.7 × 10 cm chromatographic plates from your instructor. These plates are composed of a layer of cellulose 160 μ thick supported on a polyester backing. Handle the plates by the edges only. *Never touch the adsorbent layer with your hands.* The cellulose layer is easily scratched or chipped so marking and spotting of the plates must be done carefully.

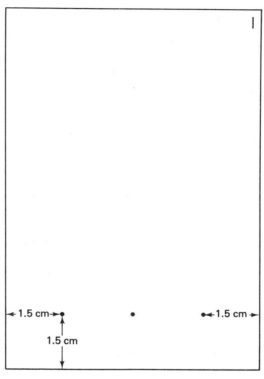

Figure 7-1.

Using a centimeter rule for measuring, make three *tiny* pencil marks on one of the narrow ends of each plate as shown in Fig. 7-1. Number each plate in the upper right-hand corner. A clean, dry 600 ml beaker will serve as a developing chamber. Fill it with sufficient (about 20 ml) n-propanol-NH_3-H_2O developing solvent (CAUTION: Avoid prolonged inhalation of vapor) to give a depth of 0.5 cm or a little less. Place a piece of aluminium foil on top of the beaker and form it to give a reasonably tight cover. Allow the beaker to stand so that it can fill with solvent vapor while you are spotting the plates. Use a 3 mm diameter glass stirring rod with a slightly rounded end as a tool for applying the spots to the chromatographic plates. Before actually spotting the plates, practice making spots on a piece of filter paper using the acetic acid known solution. Make a spot by dipping just the end of the clean, dry rod into the solution and touching it quickly and lightly to the paper. A spot 3 to 5 mm in diameter is required. When you feel that you can reliably produce spots of this size, you are ready to spot the plates. *Be sure to clean and dry the glass rod each time you use a different solution.* Choose any four of the five known individual acid solutions, the known acid mixture, and your unknown acid mixture to apply to the plates. The known mixture contains all five of the acids and your unknown mixture contains two or more acids. Apply one solution at each pencil mark. After making a spot, let it dry and repeat the procedure *two more times* at the same place with the same solution. This will apply a total of about 3 μl (microliters; 1 μl = 1 × 10^{-6} l) at each mark. In the chart in the laboratory report, record the identity of each substance and where it was spotted.

When all the spots have dried thoroughly, place both plates in the developing chamber so that the end of the plate that has been spotted is dipping into the solvent. Initially the solvent should be well below the spots but it will quickly rise up the plate by capillarity. The plates should not touch each other. Replace the foil cover tightly on the beaker; *be careful not to disturb the solvent.* The beaker must be left completely undisturbed until the development of the plates is finished.

When the solvent front has moved up the plates so that it is 0.5 to 0.75 cm from the top of each plate (about 60 to 90 minutes), remove the plates from the developing chamber and let them air-dry for a few minutes. Place them in a 60°C oven for 15 minutes to completely dry them. It is necessary that all the NH_3 from the solvent be removed; otherwise a blue background will be obtained on the plate when it is sprayed with indicator. It is a bit difficult to detect blue spots on a blue background. With a pencil, lightly mark the upper limit reached by the solvent front.

Spray the dried plates with a fine mist of bromophenol blue solution. Only a small amount is necessary to detect the spots so *do not soak the plate or even spray it until it looks damp.* The acids will show up as blue spots on a yellow background. The spots may fade rather quickly; as soon as the indicator has dried, lightly outline the position of each spot with a pencil mark.

Measurement of R_f values. Using a metric rule, measure the distance from the pencil dot indicating the location the sample was placed on the plate to the center of the spot. This is the distance that the sample has traveled. *At the same place on the chromatogram*, measure the distance from the pencil dot to the top of the solvent front. This is the distance traveled by the solvent during the same length of time as the sample was moving. Do this for each spot obtained on the chromatogram. Record these values on the Report sheet and calculate the R_f values (to two decimal places) using Eq. (1). By comparing the R_f values and positions of the known acids with those of the acids in the unknown mixture, identify the components of your unknown.

SPECIFIC REFERENCE

1. H. Bayzer, "Die Trennung von gesättigten, aliphatischen Monocarbonsäuren auf Celluloseschichten" (The Separation of Saturated Aliphatic Monocarboxylic Acids on Cellulose Layers), *J. Chromatogr.*, **27**, 104-108 (1967).

GENERAL REFERENCES

J. M. Bobbitt, A. E. Schwarting, and R. J. Gritter, "Introduction to Chromatography", Reinhold Book Corp., New York, N.Y., 1968, Chap. 3.

J. A. Dean, "Chemical Separation Methods", Von Nostrand Reinhold Company, New York, N.Y., 1969, Chap. 10.

E. Stahl, ed., "Thin-Layer Chromatography", 2nd ed., translated by M. R. F. Ashworth, Springer-Verlag New York Inc., New York, N.Y., 1967.

EXERCISES

1. A student records the data from his chromatogram: distance spot moved = 77 mm; distance solvent front moved = 126 mm. Calculate the R_f value for this spot on this chromatogram.
2. What if the student in Exercise 1 had used an inch ruler rather than a metric ruler? Would his R_f value be the same or not? If you cannot answer the question directly, try the calculation by converting his data to inches using the conversion 1 inch = 25.4 mm.
3. If a solution of an organic acid contains 4 mg of acid per milliliter and you put 3 µl of solution onto a chromatogram, how many micrograms of acid have you put on? If the molecular weight is 88.10, how many moles of acid are put on?

(Answers are on p. 245)

REPORT THIN-LAYER CHROMATOGRAPHY

1. Unknown No. _____

2. Spotting of chromatographic plates:

Plate 1		Plate 2	
Position	Substance(s)	Position	Substance(s)
Left		Left	
Middle		Middle	
Right		Right	

3. (a) Individual acids:

Acid					
Distance spot moved (mm)					
Distance solvent front moved (mm)					
R_f					

(b) Known mixture:

Acid	Acetic	Propionic	Butyric	Valeric	Caproic
Distance spot moved (mm)					
Distance solvent front moved (mm)					

(b) Known mixture (*cont.*)

R_f					

(c) Unknown mixture:

Distance spot moved (mm)					
Distance solvent front moved (mm)					
R_f					
Identity					

4. The acidic color of the bromophenol blue indicator is yellow. Why are the spots indicating the location of the acids a blue color?

Name _____ Section _____ Grade _____

Experiment 8

Ion-Exchange Chromatography

Chromatography is an important method with numerous applications for the separation of mixtures of substances. It is based upon the distribution of a solute between a stationary phase and a moving fluid phase. The distribution is expressed by the equation

$$D = \frac{[S]_s}{[S]_m}$$

where $[S]_s$ is the concentration of the solute in the stationary phase, $[S]_m$ is the concentration in the moving phase, and D is the *distribution coefficient* (actually a kind of equilibrium constant). By judiciously choosing experimental conditions so that the substances have largely differing distribution coefficients, many separations are possible.

Ion exchange is one of several types of chromatography. It involves the exchange of an ion in the solution for an ion of like charge that is bound to the stationary phase. The stationary phase may consist of a natural substance but most frequently it is composed of synthetic organic polymers called *resins*. The most commonly used resin is a copolymer of styrene and divinylbenzene to which ionic groups have been attached. If the resin is a cation-exchange resin, it will contain acidic groups such as $-COOH$ or $-SO_3H$ that are capable of exchanging a proton for another cation.

$$-COOH + Na^+ \rightleftarrows -COONa + H^+$$

$$2-SO_3H + Zn^{2+} \rightleftarrows (-SO_3)_2Zn + 2H^+$$

If the resin is an anion-exchange resin, it will contain basic groups such as amine or ammonium groups.

$$-(CH_3)_3NOH + Cl^- \rightleftarrows -(CH_3)_3NCl + OH^-$$

$$2-(CH_3)_3NCl + [CoCl_4]^{2-} \rightleftarrows [-(CH_3)_3N]_2[CoCl_4] + 2Cl^-$$

The relative binding of ions by a resin is dependent on a number of factors. (1) The higher the charge on the ion, the more strongly it is held by the resin, all other conditions being the same. (2) The nature of the ion-exchange resin affects the relative binding of ions. (3) The solvent or solution which is used as the moving phase greatly influences the exchange of an ion because it affects the types of species which can exist. For example, in a solution having a very large Cl^- concentration, Cd(II) exists as the anion $[CdCl_4]^{2-}$, while in pure water it exists as the cation $[Cd(H_2O)_6]^{2+}$. The $[CdCl_4]^{2-}$ ion would not be bound by a cation-exchange resin, while the $[Cd(H_2O)_6]^{2+}$ would be strongly bound.

Probably the most widespread application of ion-exchange chromatography is its use in the household water softener. All multiply charged cations are removed and replaced with sodium ions. Other applications include the analysis of trace amounts of metals in water, milk, and other substances. In scientific and technical fields it is used to prepare, separate, and purify ionic substances.

An efficient method of accomplishing an ion exchange is to place the resin in a piece of tubing that is constricted at the bottom. A solution is continually in contact with a fresh surface of resin as it passes down the column. The repeated adsorption and removal of an ion, or any species, by a resin while in the presence of a moving liquid is called *elution*. The liquid put into the top of a column is the *eluent* and the liquid coming out at the bottom of the column is the *eluate*. In ion-exchange chromatography, the species to be separated are adsorbed at the top of the column and are separated into bands as they move down the column because of the difference in their distribution coefficients. Ions may also be separated by adsorption on the column followed by removal of them individually by changing the eluting solvent. Both of these cases occur by an ion-exchange process. This experiment employs the latter case.

It is possible to separate each metal ion from a solution containing Fe(III), Co(II), Ni(II), Cu(II), Zn(II), and Cd(II) by cation-exchange chromatography. The separation is based upon the relative stabilities of the chloro anions of these metals. The solution that is placed on the column contains the metals in the form of their hydrated ions $[Fe(H_2O)_6]^{3+}$, $[Co(H_2O)_6]^{2+}$, $[Ni(H_2O)_6]^{2+}$, $[Cu(H_2O)_6]^{2+}$, $[Zn(H_2O)_6]^{2+}$, and $[Cd(H_2O)_6]^{2+}$. In this form they can exchange with the H^+ of the $-SO_3H$ groups present in the resin that is to be used. The hydrated ions are bound to the resin and the H^+ released into the solution. The metal ions are removed from the column one by one by eluting with 0.5 M HCl in aqueous acetone solutions. In these solutions the hydrated cations are converted to chloro anions and can no longer be adsorbed by the resin. An exception to this is Ni(II), which does not form a chloro anion under these conditions. All the chloro anions would come through the column together if it were not for the fact that the formation of the chloro anion of a given metal is dependent on the relative amounts of water and acetone in the 0.5 M HCl solution. For example, in a solution of 0.5 M HCl composed of 40% acetone and 60% water by volume the chloro anion of Cd(II) is formed but not the chloro anions of Fe(III), Co(II), Ni(II), Cu(II), and Zn(II). Therefore, Cd(II) can be separated from the others by eluting with this solvent.

You will receive an unknown solution containing two or three of the metal ions cited above and separate them by column chromatography. You will also be told which eluting solutions to use (by letter symbol, see Table 8-2) and in what order. You will be asked to test the eluate to detect the presence or absence of a metal ion and to confirm the identity of the ion.

Before beginning the elution, preliminary tests are to be done on the unknown solution in order to eliminate or confirm the possible presence of *some* of the metal ions. This will reduce the number of different tests that must be performed on the eluate and in general make your life easier during the laboratory period.

The basis for the preliminary and confirmatory tests lies in the descriptive chemistry of the six metal ions. The formulas and colors of the species encountered in these tests are listed in Table 8-1.

All six metal ions react with sulfide ion (S^{2-}) to form insoluble metal sulfides. The divalent ions react according to the general equation (where $M^{2+} = Co^{2+}$, Ni^{2+}, Cu^{2+}, Zn^{2+}, Cd^{2+}):

$$M^{2+}(aq) + S^{2-}(aq) \longrightarrow MS(s)$$

The sulfide ion is a strong enough reducing agent to reduce Fe(III) to Fe(II).

$$2\ Fe^{3+}\ (aq)\ +\ 3\ S^{2-}\ (aq)\ \longrightarrow\ 2\ FeS\ (s)\ +\ S\ (s)$$

Commonly used sources of sulfide ion are sodium sulfide, ammonium sulfide, and hydrogen sulfide. Before testing the eluate with sulfide ion, the solution is made basic because some of the metal sulfides are soluble in acid solution. The original metal ion solutions are nearly neutral so no pH adjustment is necessary.

All six of the metal ions form insoluble hydroxides when reacted with a stoichiometric amount of sodium hydroxide or other strong base. The divalent metal ions react according to the general equation (where M^{2+} is again Co^{2+}, Ni^{2+}, Cu^{2+}, Zn^{2+}, Cd^{2+}):

$$M^{2+}\ (aq)\ +\ 2\ OH^-(aq)\ \longrightarrow\ M(OH)_2\ (s)$$

Iron(III) is again different but only because it has a +3 charge.

$$Fe^{3+}\ (aq)\ +\ 3\ OH^-(aq)\ \longrightarrow\ Fe(OH)_3\ (s)$$

Zinc(II) and cadmium(II) hydroxides are amphoteric (capable of reacting with acid or base) and will dissolve in excess NaOH forming anionic species.

$$Zn(OH)_2\ (s)\ +\ 2\ OH^-(aq)\ \longrightarrow\ Zn(OH)_4^{2-}(aq)$$

$$Cd(OH)_2\ (s)\ +\ 2\ OH^-(aq)\ \longrightarrow\ Cd(OH)_4^{2-}(aq)$$

When the metal ions are reacted with aqueous ammonia (ammonium hydroxide), the hydroxides are first formed, but all of them [except $Fe(OH)_3$] will redissolve with excess ammonia forming ammine complexes. Notice that the cobalt(II) complex is oxidized by oxygen in the air to the corresponding cobalt(III) complex.

$$Zn(OH)_2\ (s)\ +\ 4\ NH_3\ (aq)\ \longrightarrow\ [Zn(NH_3)_4]^{2+}\ (aq)\ +\ 2\ OH^-(aq)$$

$$Cd(OH)_2\ (s)\ +\ 4\ NH_3\ (aq)\ \longrightarrow\ [Cd(NH_3)_4]^{2+}\ (aq)\ +\ 2\ OH^-(aq)$$

$$Cu(OH)_2\ (s)\ +\ 4\ NH_3\ (aq)\ \longrightarrow\ [Cu(NH_3)_4]^{2+}\ (aq)\ +\ 2\ OH^-(aq)$$

$$Ni(OH)_2\ (s)\ +\ 6\ NH_3\ (aq)\ \longrightarrow\ [Ni(NH_3)_6]^{2+}\ (aq)\ +\ 2\ OH^-(aq)$$

$$Fe(OH)_3\ (s)\ +\ NH_3\ (aq)\ \longrightarrow\ N.\ R.\ (no\ reaction)$$

$$Co(OH)_2\ (s)\ +\ 6\ NH_3\ (aq)\ \longrightarrow\ [Co(NH_3)_6]^{2+}\ (aq)\ +\ 2\ OH^-(aq)$$

$$4\ [Co(NH_3)_6]^{2+}\ (aq)\ +\ O_2\ (g)\ +\ 2\ H_2O\ (l)\ \longrightarrow\ 4\ [Co(NH_3)_6]^{3+}\ (aq)\ +\ 4\ OH^-(aq)$$

The six metal ions react with excess thiocyanate ion, SCN^-, to form soluble complexes, some of which are colored. In some cases it is necessary to have an organic solvent present because the complexes dissociate in water. The divalent metal ions react according to the general equation (where M^{2+} is again Co^{2+}, Ni^{2+}, Cu^{2+}, Zn^{2+}, Cd^{2+}):

$$M^{2+}\ (aq)\ +\ 4\ SCN^-(soln)\ \longrightarrow\ [M(SCN)_4]^{2-}(soln)$$

Once again Fe(III) is different and once again this difference can be attributed to its higher charge.

$$Fe^{3+}\ (aq) + 6\ SCN^-\ (aq) \longrightarrow [Fe(SCN)_6]^{3-}\ (aq)$$

The test used to confirm the presence of Ni(II) in a fraction of the eluate is the formation of the red precipitate nickel(II) dimethylglyoximate, $NiC_8H_{14}N_4O_4$. The following equation shows the structures of dimethylglyoxime and the compound it forms with nickel(II).

$$2\ \begin{array}{c} H_3C\\ \\ H_3C \end{array}\!\!\!\!\!\!\begin{array}{c} C=NOH\\ |\\ C=NOH \end{array} + Ni^{2+} + 2\,OH^- \longrightarrow \text{[Ni(dmg)}_2\text{ complex]} + 2\,H_2O$$

The test must be performed in basic solution because the nickel(II) dimethylglyoximate is soluble in acid.

The chloro anions of Fe(III), Cu(II), and Co(II) in acetone-0.5 M HCl are colored.

$[FeCl_4]^-$ and $[CuCl_4]^{2-}$ yellow

$[CoCl_4]^{2-}$ blue

The yellow color of $[CuCl_4]^{2-}$ in the eluate is used as the confirmatory test for Cu(II), providing it also gives a negative test for Fe(III) with SCN^-.

TABLE 8-1
Summary of Qualitative Tests

Aqueous Solution	Insoluble Sulfide	Insoluble Hydroxide	Solution with Excess Ammonia	Solution with Excess Thiocyanate
$[Fe(H_2O)_6]^{3+}$ yellow	FeS black	$Fe(OH)_3$ red-brown	$Fe(OH)_3$ does not dissolve	$[Fe(SCN)_6]^{3-}$ deep red
$[Co(H_2O)_6]^{2+}$ pink	CoS black	$Co(OH)_2$ blue (changes to pink)	$[Co(NH_3)_6]^{3+}$ yellow-brown to brown	$[Co(SCN)_4]^{2-}$ blue
$[Ni(H_2O)_6]^{2+}$ green	NiS black	$Ni(OH)_2$ pale green	$[Ni(NH_3)_6]^{2+}$ light blue	$[Ni(SCN)_4]^{2-}$ blue-green
$[Cu(H_2O)_6]^{2+}$ blue	CuS black	$Cu(OH)_2$ blue	$[Cu(NH_3)_4]^{2+}$ deep blue	$[Cu(SCN)_4]^{2-}$ yellow-green
$[Zn(H_2O)_6]^{2+}$ colorless	ZnS white	$Zn(OH)_2$ white	$[Zn(NH_3)_4]^{2+}$ colorless	$[Zn(SCN)_4]^{2-}$ colorless
$[Cd(H_2O)_6]^{2+}$ colorless	CdS yellow	$Cd(OH)_2$ white	$[Cd(NH_3)_4]^{2+}$ colorless	$[Cd(SCN)_4]^{2-}$ colorless

This experiment may require more than one period. It is recommended that in the period preceding the one scheduled for doing the actual separation you (1) do the preliminary tests and (2) weigh out and wash the resin through step 3 of the washing procedure. Then store the resin in distilled water in a stoppered flask or test tube.

PROCEDURE

CAUTION: Acetone and the acetone-containing solutions used in this experiment are highly flammable.

Preliminary testing. The following qualitative tests should be made on your unknown solution. *For each test*, place 10 drops of the unknown in a clean test tube. For tests 1 to 3, dilute to approximately 1 ml with distilled *water*; for test 4, dilute to 1 ml with *acetone*. Record the results on the Report sheet and decide which metal ions are likely to be present and absent in the unknown.

Test 1. Add 3 drops of 2 M Na_2S, shaking or stirring and observing the color of the precipitate *after each drop*. The presence of a dark-colored precipitate may mask a light-colored precipitate.
Test 2. Add 4 drops of 2 M NaOH and mix.
Test 3. Add 3 drops of 15 M NH_3 and mix. If a precipitate is present, centrifuge or allow it to settle and observe the colors of the precipitate and the supernatent solution.
Test 4. Add 4 drops of 1 M NH_4SCN and mix.

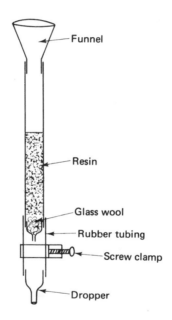

Figure 8-1. Chromatography column.

Preparation of the column and resin. Obtain a plastic chromatography column that has an inside diameter of 7 mm. Place a small piece of glass wool in the bottom end of the column to support the resin in the column and push it down in with a stirring rod if your column does not disassemble at the bottom. Place a small piece (5 to 6 cm long) of rubber tubing firmly on the bottom end of the column and put a medicine dropper* in the other end of the tubing. A screw clamp placed on the tubing between the column and the dropper will serve as a means of regulating the flow of eluate from the column. Clean all your small test tubes and number them consecutively.

Weigh 2.5 g of a 50 to 100-mesh sulfonic acid-type cation-exchange resin (Dowex 50W X8) into a 100 ml or 150 ml beaker. Wash the resin with the solvents or solutions in the

---

*To make a more compact apparatus, cut off the medicine dropper 1.5 cm from the beginning of the tapered part and fire-polish the cut end.

order listed below. For each washing, add the specified quantity of wash liquid, stir the resin, and allow it to settle. Decant (see p. 36) the supernatant liquid and any fine particles floating on the surface into a beaker. Never discard any resin. It can be washed and reused.

1. 15 ml of distilled H_2O
2. 15 ml of 3 M HCl, two times
3. 15 ml of distilled H_2O
4. 10 ml of the first solvent that you will use for eluting (see Table 8-2)

TABLE 8-2

Eluting Solvents

Solvent	Composition
A	40% acetone--0.5M HCl
B	60% acetone--0.5M HCl
C	75% acetone--0.5M HCl
D	90% acetone--0.5M HCl
E	3M HCl

Place another 10 ml of the first eluting solvent in the beaker with the resin and allow the resin to soak at least 5 minutes. Swirl the resin and carefully pour the mixture into the column, allowing the solvent to pass through the column at the maximum rate. Air bubbles trapped in the column reduce the efficiency of the separation and should be eliminated by tapping the column while the resin is settling. Add resin until the height of the resin column is 12 cm ± 0.5 cm. *Never let the resin become dry at any time until the experiment is completed* and always add solvents to the column carefully. When *almost* all the solvent has soaked into the resin, add another 10 ml of your first eluting solvent and adjust the screw clamp so that the solvent flows through the column at a rate of $\frac{2}{3}$ to 1 ml per minute. The flow rate can be readjusted at any time during the course of the separation. Remember, it is better to have too slow a flow rate than too fast because a better separation is achieved with a slower rate. If at any time you get behind in your testing or you need to wash test tubes, you can stop the solvent flow altogether for a few minutes until you are ready to proceed once again.

Separation of a metal ion mixture. When *almost* all the solvent has soaked into the resin, add 5 drops of your unknown metal ion solution followed by 10 drops of your first eluting solvent. When almost all this has soaked into the resin, repeat with another 10 drops followed by 10 ml of the first eluting solvent. Start collecting the eluate immediately in 2 ml portions in small test tubes. A convenient guide for estimating this volume can be made by placing 2 ml of water into another test tube of the same size. Test each 2 ml fraction according to the tests given in Table 8-3 and record the results on the Report sheet. Once a metal ion in the eluate has been positively identified, perform only the confirmatory test on succeeding fractions of eluate. Keep eluting with the same solvent until a negative test is obtained for the first ion to come off the column. Add more solvent (in 4 ml increments) if necessary, but a total of 26 ml of a given solvent should suffice.

Once a negative test has been obtained, allow the remainder of the first solvent to *almost* soak into the resin and add 10 ml of the second solvent. Continue collecting 2 ml fractions and testing them for the presence of metal ions. Add more solvent to the column if necessary to completely elute the second metal ion.

If a third solvent is given on your solvent list, add 10 ml of it to the column when you have finished eluting with the second solvent. Proceed in the same manner as you did for the first and second solvents.

When all the metal ions have been eluted, allow 10 ml of 3 M HCl to pass through the column at the maximum elution rate and follow it with 10 ml of distilled water. Empty the resin into the container provided in the laboratory and discard the glass wool.

TABLE 8-3

Confirmatory Tests

A. Na_2S test	B. NH_4SCN test	C. Dimethylglyoxime test
To 1 ml of eluate, add 6M NaOH until basic to litmus (2 drops if the eluting solvent is acetone--0.5M HCl or 5 drops if the solvent is 3M HCl), add 2 drops of 2M Na_2S, and mix (a) A precipitate of CdS or ZnS is positive confirmatory evidence for the presence of Cd(II) or Zn(II); no further tests are required (b) If a black precipitate is obtained, continue with test B (c) Absence of a precipitate indicates that no metal ions are present in the eluate	A yellow-colored eluate indicates either Fe(III) or Cu(II); do steps (a) and (b); a colorless eluate indicates either Co(II) or Ni(II); do step (c) (a) To 10 drops of eluate diluted to 1 ml, add 1 drop of 1M NH_4SCN and mix; a red color confirms the presence of Fe(III) and no further tests are required* (b) A yellow-colored eluate that gives a negative test for Fe(III) with NH_4SCN is positive confirmation of the presence of Cu(II); no further tests are required (c) To 10 drops of eluate, add 3 drops of 1M NH_4SCN, dilute to 2 ml with acetone, and mix; a blue color confirms the presence of Co(II); no further tests needed (d) If tests for Fe(III), Cu(II), and Co(II) are negative, continue with test C	To 10 drops of eluate, add 5 drops of 15M NH_3 and 2 drops of dimethylglyoxime and mix (a) A red precipitate confirms the presence of Ni(II)

*The confirmatory test for iron(III) is quite sensitive and iron(III) impurities are quite common. Therefore a very faint test for Fe(III) may be considered as a negative test.

SPECIFIC REFERENCE

1. J. S. Fritz and T. A. Rettig, "Separation of Metals by Cation Exchange in Acetone-Water-Hydrochloric Acid", *Anal. Chem.*, 34, 1562 (1962).

GENERAL REFERENCES

J. M. Bobbitt, A. E. Schwarting, and R. J. Gritter, "Introduction to Chromatography", Reinhold Book Corp., New York, N.Y., 1968, Chap. 1 and 4.

I. M. Kolthoff, E. B. Sandell, E. J. Meehan, and S. Bruckenstein, "Quantitative Chemical Analysis", 4th ed., The Macmillan Company, New York, N.Y., 1969, Chap. 13.

R. L. Pecsok and L. D. Shields, "Modern Methods of Chemical Analysis", John Wiley and Sons, Inc., New York, N.Y., 1968, Chap. 4 and 6.

C. H. Sorum, "Introduction to Semimicro Qualitative Analysis", 4th ed., Prentice-Hall, Inc., Englewood Cliffs, N.J., 1967, Chap. 4 and 5.

EXERCISES

1. A student is given an unknown solution that contains two ions. He performs the preliminary tests and obtains the following results.

    Color of solution: violet
    With Na_2S: black precipitate

     With NaOH: blue precipitate
     With NH_3: blue solution turning green
     With NH_4SCN and acetone: yellow-green solution

  What were the ions in the solution as judged from these tests?
2. A student was testing the eluate from his column and obtained the following results.

     Color of eluate: yellow
     With Na_2S: black precipitate
     With NH_4SCN: no red color

  What ion was present in the eluate?
3. Another student was testing the eluate from his column and obtained the following results.

     Color of eluate: colorless
     With Na_2S: black precipitate
     With NH_4SCN: no red color
     With NH_4SCN and acetone: blue color

  What ion was present in the eluate?
4. In adding eluent to his column, a student accidently added 2 ml of 40% acetone-0.5 M HCl when he should have added 2 ml of 90% acetone-0.5 M HCl. What should he do now?
5. Another student accidently added 10 ml of 90% acetone-0.5 M HCl when he should have added 40% acetone-0.5 M HCl. What should he do now?

(Answers are on p. 246)

REPORT ION-EXCHANGE CHROMATOGRAPHY

1. Unknown No. _____

    Eluting solvent: first _____, second _____, third _____

2. Preliminary tests:

    (a) Color of unknown solution _____

    (b)

Test No.	Precipitate? (yes or no)	Results (color, etc.)
1		
2		
3		
4	No	

    (c) Ions judged definitely present _____

    Ions judged definitely absent _____

    Other ions possibly present _____

3. Elution and confirmatory tests:

Eluate Number	Eluent	Color of eluate	Test used for confirmation (A, B, or C)	Results	Ion present
1					
2					
3					
4					
5					
6					
7					
8					
9					
10					
11					
12					

Elute Number	Eluent	Color of eluate	Test used for confirmation (A, B, or C)	Results	Ion present
13					
14					
15					
16					
17					
18					
19					
20					
21					
22					
23					
24					
25					
26					
27					
28					
29					
30					
31					
32					
33					
34					
35					

4. Ions found in unknown _____

5. What type of problems did you encounter? What might you have done differently to overcome these problems?

6. If by accident you added 10 drops of the metal ion unknown solution instead of 5 drops to the column, how would this affect the time and sharpness of the separation?

7. If you wanted to separate a large quantity of metal ion solution, how might you modify the experiment in order to do this?

Name _____ Section _____ Grade _____

Experiment 9
Chemical Mystery Theater

The scene is a chemistry laboratory. Standing by the laboratory bench is a general chemistry student with N bottles before him. The bottles are numbered 1 through N. He is puzzled and perplexed. Suddenly a forbidding figure appears behind him (his lab instructor) who silently hands him a piece of paper with the names of N chemicals written on it. With a devilish chuckle the figure says "That is what is in your bottles. Find out which bottle contains which chemical." The student, after a moment's reflection, pictures himself as a combination Sherlock Holmes, Dick Tracy, and Batman ready to observe even the most minute detail of the reactions of the chemicals in the bottles and solve the "Great Mystery of the Unlabeled Bottles." He goes to work. His excitement can barely be contained as he begins mixing a little of solution 1 with a little of solution 5, a little of solution 2 with solution 4, etc., observing a precipitate here, a gas there, and occasionally a color change. More often than not he observes nothing at all. Finally he shouts, "EUREKA, I've done it," and he feverishly writes his conclusions and turns them in to his lab instructor.

PROCEDURE

At the beginning of the lab period, your instructor will supply you with N bottles, each of which contains an aqueous solution (one may just contain pure water). The bottles will all contain solutions that you have utilized in previous experiments. You will also be given a list of what those N solutions are. Your job is to determine which chemical is in which bottle using only those N chemicals and the equipment you have in your drawer.

You may bring any written material you want with you to lab. As a minimum you should bring your text, lab manual, and class notes. Keep in mind that you have only a limited amount of each solution, so carry out the reactions on a small scale — a few drops of each solution in a test tube.

EXERCISES

1. A chemist walks into his laboratory one morning and finds four labels laying on the lab bench near four bottles without labels. His budget is limited, as evidenced by the cheap labels he is using, and so he wants to identify the contents of the bottles and relabel them rather than discard them and prepare fresh solutions. He designates the bottles 1,

2, 3, and 4 and notes that the labels he found read 1 *M* $AgNO_3$, 6 *M* NH_4OH, 6 *M* NaCl, and 6 *M* HCl. The following observations were noted.

1 Ag
2 Na
3 NH
4 HCL

Solution 1 added to solution 2: white precipitate
Solution 1 added to solution 4: white precipitate
Solution 3 added to white precipitate formed in either test 1 or test 2 above: precipitate dissolved
Solution 4 is acidic to litmus

Which label goes on which bottle?

2. A student in chemistry laboratory, getting ready to do an experiment, obtained solutions of K_2CrO_4, HCl, Na_2CO_3, $Mn(NO_3)_2$, NaOH, H_2S, and KBr. About that time his girlfriend appeared at the lab door. Since she did not have her safety glasses, he went out into the corridor to talk to her. His neighbor down the bench was known to be a goof-off and so when the student came back into the lab he checked his solutions. Sure enough, labels had been switched — he could tell because the yellow solution was labeled HCl. Calling the bottles *A*, *B*, *C*, *D*, *E*, *F*, and *G* for temporary identification, he recorded the following observations.

Solution *D* was yellow and solution *C* didn't quite appear colorless.
Solution *B* had a terrible smell.
Solution *D* turned orange when solution *E* was added to it.
Solution *G* gave off a gas (seemed to be odorless) when solution *E* was added to it.
Solution *B* added to solution *C* gave a pink precipitate.
Solution *F* also gave a pink precipitate when added to solution *C* but the pink quickly discolored.

Which label goes on which bottle?

(Answers are on p. 246)

A KBr
B H_2S
C NaOH Mn(NO)$_3$
D K_2CrO_4
E HCl
F NaOH
G Na_2CO_3

REPORT CHEMICAL MYSTERY THEATER

1. In a systematic fashion, write your observations as you carry out the reactions. You may also want to speculate on the meaning of those observations and so space is provided here for both observation and speculation and an opportunity to keep the fact separate from the "fiction."

Procedure	Observation	Possible conclusions

2. Write balanced net ionic equations for each of the reactions described in No. 1.

3. Summary of conclusions:

Bottle number	Chemical	Bottle number	Chemical
_____	_____	_____	_____
_____	_____	_____	_____
_____	_____	_____	_____
_____	_____	_____	_____
_____	_____	_____	_____
_____	_____	_____	_____

Name _____ Section _____ Grade _____

Experiment 10
Synthesis of a Coordination Compound

A coordination compound is a compound, or ion within a compound, formed from a Lewis acid and one or more Lewis bases. The Lewis acid, called the *acceptor*, is a metal atom or ion and the Lewis base, called the *donor* or *ligand*, may be an atom, ion, or neutral molecule.

Late in the nineteenth century, a chemist by the name of A. Werner first proposed a theory regarding the structure of coordination compounds. Werner's theory states that a definite number of ligands are arranged around a central atom in a simple, spatial, geometric pattern. Although the methods used to study these compounds have increased in sophistication, Werner's theory has been verified. Today, thousands of coordination compounds have been synthesized but, compared to the possibilities, they are like a thimbleful of water removed from a bucketful.

The coordination number describes the number of donor atoms or groups that surround a central atom or ion. In the past it was thought that all metal ions exhibited a coordination number of 4 or 6 but compounds are now known that have coordination numbers from 2 to 12. A majority of coordination compounds do, however, have a coordination number of 4 or 6. Some metal ions can exhibit more than one coordination number, depending on the conditions and the ligand(s). For example, Co(II) is known to form compounds in which its coordination number is 4, 5, or 6. In general, the coordination number is determined by the size and electronic configuration of the central atom, the donor strength of the ligand, and the size and shape of the ligand.

Any atom possessing a nonbonding pair of electrons is a potential donor. Thus molecules that contain oxygen, nitrogen, sulfur, phosphorus, or arsenic (the last two in their lowest oxidation states) are generally good ligands. Halide ions are also donors.

Some ligands contain more than one donor atom per molecule and are called *polydentate* ligands. A ligand that contains one donor atom per molecule is referred to as a monodentate (having one tooth); two donor atoms per molecule, bidentate; three donor atoms per molecule, tridentate; etc. If all donor atoms of a polydentate ligand are bonded to the same central metal ion, the ligand is called a *chelating agent* and the resulting coordination compound is a *chelate*.

For a more thorough discussion of the topic, you should consult your text and the books listed under General References at the end of this section.

This experiment gives the directions for the synthesis of coordination compounds having coordination numbers of 5, 6, and 7.

GENERAL REFERENCES

J. V. Quagliano and L. M. Vallarino, "Coordination Chemistry", D. C. Heath and Co., Lexington, Mass., 1969.

F. Basolo and R. C. Johnson, "Coordination Chemistry", W. A. Benjamin, Inc., New York, N.Y., 1964.

M. M. Jones, "Elementary Coordination Chemistry", Prentice-Hall, Inc., Englewood Cliffs, N.J., 1964.

R. K. Murmann, "Inorganic Complex Compounds", Reinhold Publishing Corp., New York, N.Y., 1964.

D. F. Martin and B. B. Martin, "Coordination Compounds", McGraw-Hill Book Co., New York, N.Y., 1964.

I. Zinc Acetylacetonate Monohydrate — Coordination Number 5

Acetylacetone (correctly named 2,4-pentanedione) functions as a bidentate ligand and the two keto oxygen atoms coordinate to a metal ion.

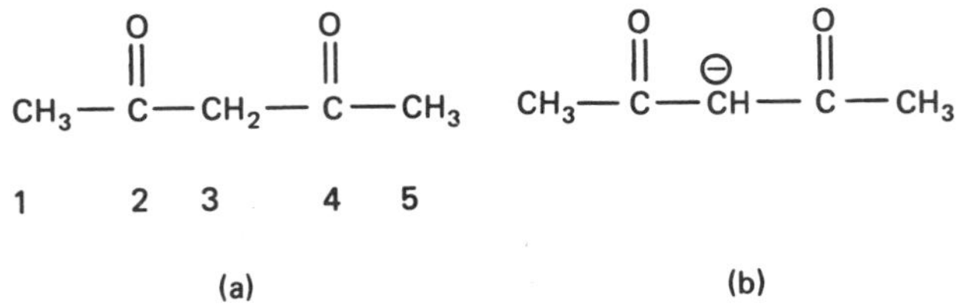

Figure 10-1. (a) Acetylacetone, (b) Acetylacetonate ion.

When it is reacted with a base, one of the hydrogens bound to carbon No. 3 is removed as a proton, resulting in the formation of an acetylacetone anion. The structures are shown in Fig. 10-1. When the anion reacts with a metal ion, the negative charge becomes delocalized throughout the chelate ring that is formed. In Fig. 10-2, the dotted ring represents the delocalized pair of electrons.

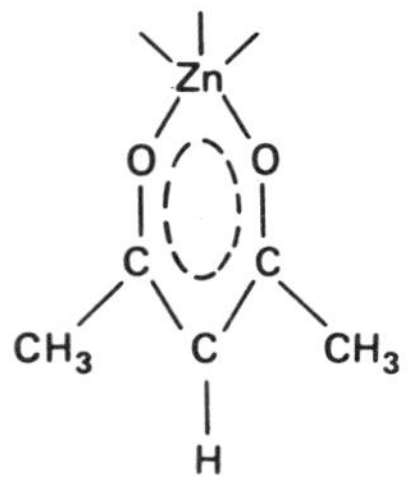

Figure 10-2. Chelated acetylacetonate ion.

There are two principal geometries that a compound having a coordination number of 5 may possess. They are trigonal bipyramidal [Fig. 10-3(a)] and square pyramidal [Fig. 10-3(b)].

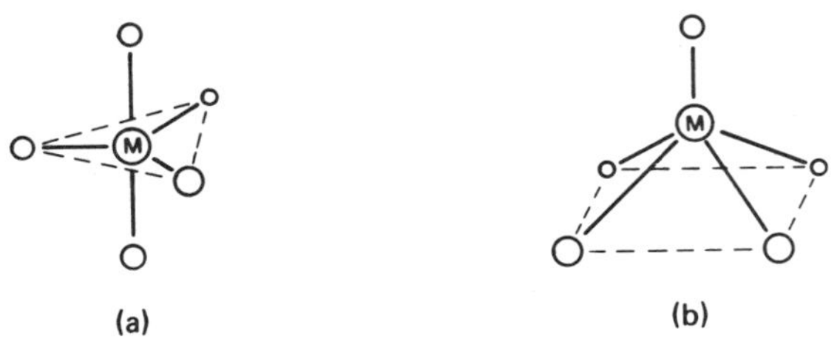

Figure 10-3. Geometries for coordination number 5.

Zinc acetylacetonate monohydrate, $Zn(C_5H_7O_2)_2 \cdot H_2O$, has the structure (a) with the water of hydration and one oxygen from each acetylacetone located in the equatorial plane.[1] The apical positions are occupied by the second oxygen of each acetylacetone.

$$(1) \quad ZnO\ (s) + 2CH_3\overset{O}{\overset{\|}{C}}CH_2\overset{O}{\overset{\|}{C}}CH_3\ (aq) \longrightarrow Zn(CH_3\overset{O}{\overset{\|}{C}}CHC\overset{O}{\overset{\|}{C}}H_3)_2 \cdot XH_2O\ (s) + H_2O\ (l)$$

$$(2) \quad Zn(CH_3\overset{O}{\overset{\|}{C}}CHC\overset{O}{\overset{\|}{C}}H_3)_2 \cdot XH_2O\ (s) \xrightarrow[\text{acetone-water}]{\text{recrystallize}} Zn(CH_3\overset{O}{\overset{\|}{C}}CHC\overset{O}{\overset{\|}{C}}H_3)_2 \cdot H_2O\ (s)$$

In this preparation the oxide is the base that reacts with the acidic hydrogen of acetylacetone. Reaction (1) is an example of a heterogeneous reaction; a solid (ZnO) is reacting with a species in solution (acetylacetone in water).

PROCEDURE

Acetone is a *highly flammable* organic solvent and must not be used or kept near an open flame. The 75% acetone solution that is used for recrystallization is also flammable. You should avoid contact with the skin or inhalation of the vapors of acetone and acetylacetone.

The experiment will require one period plus a little time during a second period. Weigh 1.6 g of zinc oxide and place it in a beaker with 50 ml of distilled water. Heat the mixture to boiling and then remove from the heat. Add 6 ml of acetylacetone dropwise with continuous stirring to the ZnO mixture. Did you observe any changes? When the addition is complete, place the beaker in an ice bath until its contents are thoroughly cooled. Suction filter the precipitate on a glass filter crucible or a small Buchner funnel with filter paper. Wash the precipitate with two or three small portions of cold water and suck out as much of the water as possible.

The solid that has been obtained is a probable mixture of $Zn(C_5H_7O_2)_2 \cdot H_2O$, $Zn(C_5H_7O_2)_2 \cdot 2H_2O$, and unreacted ZnO. It is necessary to recrystallize it in order to obtain pure $Zn(C_5H_7O_2)_2 \cdot H_2O$. To do this, prepare a solution of acetone and distilled water that contains 75 ml of acetone plus 25 ml of water and add 1 to 2 ml of acetylacetone to it. The acetylacetone serves to prevent decomposition of the product. Finally, add the precipitate to the solution and stir thoroughly. What is the appearance of the solution? Fit a long- or short-stem funnel with a piece of fine filter paper for a gravity filtration (see p. 36 on filtration techniques) and filter the mixture. The filtrate, which should be clear but not necessarily colorless, is collected in a beaker. Cover the beaker with a watch glass or a loosely fitting piece of aluminium foil and store it as directed by your instructor until the next period. It is desirable that partial evaporation of the solvent occur during the interim.

Hopefully, crystals will have formed during the time in which the beaker of solution sat between periods. Cool the crystals and remaining solution by placing the beaker in an ice bath* and, concurrently, cool about 30 ml of acetone to be used as a wash solvent. Suction filter the crystals and wash with two small portions (10 and 15 ml) of cold acetone to remove impurities, mainly the acetylacetone that was added during the recrystallization. The product should be air-dried. When it is dry, place it in a tared and labeled bottle and weigh it. Determine the melting point. Directions for analysis of the product for zinc content are given in Experiment 20.

*If all the solvent has evaporated, transfer the crystals to the filter funnel and wash with *three* portions of cold acetone.

11 Sulfatobis(ethylenediamine)copper(II) — Coordination Number 6

Ethylenediamine, $H_2NCH_2CH_2NH_2$ — abbreviated "en", is a good coordinating agent for transition metal ions and it functions as a bidentate, chelating ligand.

Copper(II) reacts with ethylenediamine to give a series of coordination compounds containing 1, 2, or 3 moles of ethylenediamine per mole of copper.[2] The product obtained is determined by the quantities of reactants used. Thus, it is important that the specified quantities of chemicals be weighed or measured with care.

Compounds that exhibit a coordination number of 6 have an octahedral or distorted octahedral structure.

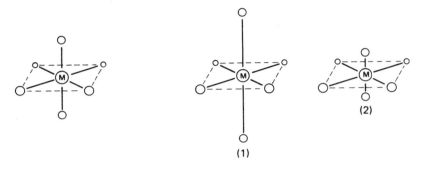

(a) Regular octahedron   (b) Distorted octahedra

Figure 10-4.

There are all degrees of distortion from the regular octahedral structure but (1), as shown in Fig. 10-4, is the most common type of distortion. A number of complexes having the formula $[Cu(en)_2X_2]$, where X equals a monovalent anion or X_2 equals a divalent anion, have been found by X-ray diffraction measurements to possess structure (1) in Fig. 10-4.

The structure of $[Cu(en)_2SO_4]$ has not been determined by X-ray crystallography as yet but it has been inferred to be the same by comparison of its infrared and visible spectra with those of $[Cu(en)_2X_2]$ compounds of known structure.[3] The sulfate groups are located at the apical positions of the octahedron and are weakly coordinated to the Cu(II). Each sulfate acts as a bidentate but nonchelating ligand and as a bridging group between two $Cu(en)_2^{2+}$ units. Therefore $[Cu(en)_2SO_4]$ is really a coordination polymer.

Figure 10-5. The structure of $[Cu(en)_2SO_4]$.

PROCEDURE

Weigh 5.0 g of copper sulfate pentahydrate, $CuSO_4 \cdot 5H_2O$, and place it in a 250 ml beaker. Dissolve it in 15 ml of distilled H_2O. It may be necessary to warm this slightly to speed up the dissolution of the $CuSO_4$. In a small graduated cylinder, carefully measure out 3.0 ml of ethylenediamine. This should be added dropwise to the $CuSO_4$ solution, stirring throughout the addition. Note your observations on the Report sheet. Add 100 ml of absolute ethanol to the final solution and cool it by placing the beaker in an ice or cold water bath until it is thoroughly chilled. The function of the absolute ethanol is to reduce the solubility of the $[Cu(en)_2SO_4]$, causing it to precipitate from solution. The crystals are to be collected in a Buchner funnel using suction filtration. Wash the crystals with two 15 ml portions of absolute ethanol and air-dry them for several minutes, breaking up lumps with your spatula. Place the $[Cu(en)_2SO_4]$ in a tared bottle and weigh it.

Before analyzing (Experiment 20) the $[Cu(en)_2SO_4]$, it should be dried for $\frac{1}{2}$ hour at 110°C.

III $Mn_3(HEDTA)_2 \cdot 10H_2O$ — Coordination Number 7

Ethylenediaminetetraacetic acid forms complexes with a great many metal ions. (See Experiment 20 for the chemistry of H_4EDTA.)

```
HOOCCH2         CH2COOH              -OOCCH2         CH2COO-
       \       /                            \       /
        NCH2CH2N                             NCH2CH2N
       /       \                            /       \
HOOCCH2         CH2COOH              -OOCCH2         CH2COOH
```

ethylenediaminetetraacetic acid

H_4 EDTA   HEDTA^{3-}

It reacts with Mn(II) to form a crystalline compound having the formula $Mn_3(HEDTA)_2 \cdot 10H_2O$ in which two of the manganese ions have a coordination number of 7. The structure of the compound was determined by X-ray crystallography.[4] It was found that the ethylenediaminetetraacetate behaves as a hexadentate (having six bonding positions) ligand with four oxygens from the carboxylate groups and the two nitrogen atoms all bound to a single Mn^{2+} ion; the seventh coordination position is occupied by a water molecule. This is depicted in Fig. 10-6.

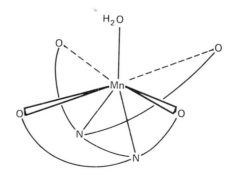

Figure 10-6. Model of the seven-coordinate manganese.

The two manganese(II) centers that have seven coordination are found in the crystal as anions, $[Mn(HEDTA)(H_2O)]^-$, linked together into infinite chains by hydrogen bonding through the hydrogen atom of HEDTA. The third manganese(II) center has six coordination and an octahedral geometry.

$Mn_3(HEDTA)_2 \cdot 10H_2O$ is synthesized by an acid-base reaction.

$$3\ MnCO_3 + 2\ H_4EDTA + 7\ H_2O \xrightarrow{\Delta} Mn_3(HEDTA)_2 \cdot 10\ H_2O + 3\ CO_2$$

PROCEDURE

This reaction is an example of a heterogeneous reaction; i.e., one reactant is in the solid state and the other reactant is in solution. The chemicals should be weighed carefully as it is desirable that they be present in stoichiometric quantities; however, it is not necessary to use an analytical balance.

Weigh 6.0 g of ethylenediaminetetraacetic acid, H_4EDTA; place it in a 400 ml beaker; and add 150 ml of distilled water. Warm the mixture until it is hot but not boiling and then remove it from the heat. The H_4EDTA will partially dissolve. Add 3.6 g of manganese(II) carbonate, $MnCO_3$, to the hot H_4EDTA mixture *in small portions* while continuously stirring. Rapid gas evolution will occur. What is the gas that is being evolved?

When the gas evolution has stopped, another 100 ml of distilled water should be added and the mixture heated below boiling to complete the reaction. Note the color of the solution and how it changes during heating. The reaction is complete when everything has gone into solution. A small amount of unreacted H_4EDTA may remain on the bottom of the beaker. In this case, suction filter the *hot* solution through a Buchner funnel with filter paper or through a glass filter crucible to remove the unreacted material. Rinse out the beaker with distilled water and return the filtrate to the beaker. Evaporate 50 to 75 ml of water from the solution. To crystallize the $Mn_3(HEDTA)_2 \cdot 10 H_2O$, cool the solution thoroughly by placing the beaker in an ice-water bath. At the same time, cool about 25 ml of distilled water to be used to wash the crystals after filtration.

The pale pink crystals of $Mn_3(HEDTA)_2 \cdot 10 H_2O$ should be filtered by suction, washed with one or two small portions of cold distilled water, and sucked as dry as possible. To dry the crystals completely, transfer them to a watch glass or small beaker and let them air-dry. When they are dry, place them in a tared sample bottle, and label the bottle with your name and the name of the product.

SPECIFIC REFERENCES

1. E. L. Lippert and M. R. Truter, *J. Chem. Soc.*, 4996 (1960).
2. G. Gordon and R. K. Birdwhistell, *J. Amer. Chem. Soc.*, 81, 3567 (1959).
3. I. M. Proctor, B. J. Hathaway, and P. Nicholls, *J. Chem. Soc. A*, 1678 (1968).
4. S. Richards, B. Pederson, J. V. Silverton, and J. L. Hoard, *Inorg. Chem.*, 3, 27 (1964).

EXERCISES

The exercises for Experiment 11 are also pertinent to this experiment.

REPORT SYNTHESIS OF A COORDINATION COMPOUND

1. What coordination compound(s) did you synthesize?

2. Describe the nature of your product (shape, color, size, melting point, etc.).

3. Describe any changes that you observed during the addition of one reagent to the other. Include such observations as color changes, viscosity change, temperature change, precipitate formation, and gas evolution, if they are pertinent to your experiment. Be specific.

4. Actual yield of coordination compound.

    Weight of bottle and compound _____

    Weight of bottle _____

    Weight of compound _____

5. Calculate the theoretical yield for your compound. (The ethylenediamine and the acetylacetone were used in excess.)

6. Calculate the percentage yield of your compound.

Experiment 11
Synthesis of Tin(IV) Iodide

Tin(IV) iodide belongs to the very important and very large class of compounds known as the *halides*. With a few notable exceptions, such as some of the rare gases, all the elements of the periodic table form halides. Their properties vary from those of the ionic, high melting, polymeric compounds, such as sodium chloride, to those of the covalent, low melting, monomeric compounds, such as ClF (b.p. = $-100.8°C$). The ionic halides can be crystallized from water, while many of the covalent halides undergo decomposition in aqueous solution.

A variety of methods for the preparation of anhydrous halides exists and the one chosen is dependent on the properties of the desired compound. The synthetic procedures can be classified according to the following general methods.

1. *Dehydration of a hydrate.* The halides that crystallize from water in the form of a hydrated salt can be dehydrated under the proper conditions.

$$BaCl_2 \cdot 2H_2O \ (s) \xrightarrow[\Delta]{120°C} BaCl_2 \ (s) + 2 H_2O \ (g)$$

2. *Direct synthesis.* The elements are reacted either in the presence or in the absence of an inert solvent. This is the most frequently used method to prepare the covalent halides. Most of the time the compound formed is the highest known halide which exists for the particular metal and halogen being reacted.

$$Xe \ (g) + \text{excess } F_2 \ (g) \xrightarrow[50 \text{ atm}]{200\text{-}250°C} XeF_6 \ (s)$$

$$V \ (s) + 2 \ Cl_2 \ (g) \xrightarrow{heat} VCl_4 \ (l)$$

Vanadium is known to have an oxidation state of +5 in some compounds, but the highest known vanadium chloride is VCl_4 and it is this compound which is obtained by direct synthesis.

3. *Decomposition of a metal oxide.* Since most metal oxides are readily available, they are a convenient starting material for the preparation of the corresponding halides.

$$ZrO_2 \ (s) + 2 \ CCl_4 \ (g) \xrightarrow{450°C} ZrCl_4 \ (s) + 2 \ COCl_2 \ (g)$$

4. *Substitution reactions.* The substitution of one halide for another is frequently used to synthesize the fluorides. Besides F_2, ClF_3 is a good fluorinating agent.

$$NiCl_2\ (s)\ +\ ClF_3\ (g)\ \xrightarrow{250°C}\ NiF_2\ (s)\ +\ Cl_2\ (g)\ +\ ClF\ (g)$$

5. *Precipitation from solution.* Some halides can be precipitated directly from aqueous solution without forming hydrates. These include the halides of the alkali metals plus those halides that are insoluble in water.

$$AgNO_3\ (aq)\ +\ HCl\ (aq)\ \longrightarrow\ AgCl\ (s)\ +\ HNO_3\ (aq)$$

Tin(IV) iodide is an example of a covalent halide that is most easily synthesized directly from its elements. The reaction is conveniently carried out in the presence of a nonreactive organic solvent, which serves the dual purpose of dissolving the I_2 and the SnI_4. The reaction is described by the following equation:

$$Sn\ (s)\ +\ 2\ I_2\ (soln)\ \xrightarrow{CCl_4}\ SnI_4\ (soln)$$

Carbon tetrachloride is chosen as the solvent because of its nonflammability and its convenient boiling point but a number of other solvents will work equally well. The SnI_4 is isolated as red-orange crystals by cooling the CCl_4 solution, whereupon it crystallizes out because of its reduced solubility.

Tin(IV) iodide is monomeric in the vapor and solid states with the four iodine atoms placed around the tin atom in a tetrahedral geometry. It is hydrolyzed by water and must be protected from contact with moisture. All glassware and apparatus that come in contact with SnI_4 or its solutions must be completely dry to avoid decomposition. The solid may be dried in air of moderate humidity, however, without being attacked.

It is of some interest to compare the properties of the tin(IV) halides (see Table 11-1). The compounds increase in covalent character in the order $SnF_4\ <\ SnCl_4\ <\ SnBr_4\ <\ SnI_4$. Tin(IV) fluoride has considerable ionic character, which accounts for its high temperature of sublimation.

TABLE 11-1
Properties of the Tin (IV) Halides

Compound	Melting Point (°C)	Boiling Point (°C)
SnF_4, white solid	705 sublime	
$SnCl_4$, colorless liquid	−33	114
$SnBr_4$, white solid	31	202 (734 torr)
SnI_4, red-orange crystals	144.5	364.5

PROCEDURE

It is desirable but not necessary to carry out this reaction in a hood. Heating of open containers of CCl_4 MUST be done in a hood, however, because the VAPORS ARE TOXIC.

Set up the apparatus as shown in Fig. 11-1. The flask is a 100 ml round-bottom or a 125 ml Erlenmeyer flask. The flask and condenser must be absolutely dry before the chemicals are introduced. Bore a cork stopper of the proper size to fit your flask and condenser. Attach rubber tubing to the inlet and outlet of the condenser and turn on the water so that it flows through the condenser at a moderate rate.

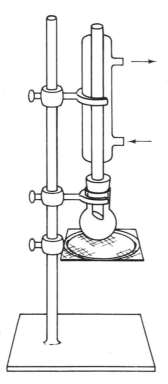

Figure 11-1. Apparatus for the preparation of tin(IV) iodide.

The following chemicals are to be weighed and introduced into the flask.

1. 3.0 g granular tin metal
2. 10.0 g iodine crystals, I_2 (causes burns on contact with the skin)
3. 20 ml carbon tetrachloride, CCl_4 (avoid contact with the skin)

Attach the flask to the condenser and check the flow rate of the cooling water. Place a wire gauze under the flask and heat gently until the reaction mixture begins to boil. The reaction is exothermic and will almost proceed spontaneously once it has been initiated. At the beginning, periodic heating will be required to maintain boiling but near the end of the reaction continuous heating will be needed. The carbon tetrachloride vapor (note the color) will be cooled and will condense on the lower third of the condenser and drip into the reaction flask. This phenomenon is called *refluxing*. The reaction is complete when the drip off the condenser is colorless (30 to 45 minutes). Record the color of the reaction mixture at the beginning and at the end of the reaction and any other pertinent observations. Is there any solid present in the flask at the end of the reaction?

Allow the flask to cool for a few minutes and while you are waiting, set up a dry filter flask and Buchner funnel for a suction filtration. (See p. 37 for suction filtration technique.) You would be wise to place a trap between the water aspirator and the filter flask. If H_2O backs up into the filter flask while the product solution is there, it's "curtains" for the SnI_4.

The condenser is removed and an additional 20 ml of CCl_4 is added to the flask. Heat this to boiling (IN THE HOOD!) and quickly filter the hot mixture to remove excess starting material. Which reagent is in excess? Add another 10 ml portion of CCl_4 to the reaction flask and again heat to boiling. Filter this through the same filter so that all the solution is together. The SnI_4 solution and any precipitated crystals are transferred to a 250 ml beaker with as little loss as possible. The beaker should be covered with a watch glass and chilled thoroughly in an ice bath, remembering that SnI_4 is moisture sensitive. The SnI_4 crystals are collected in a Buchner funnel by suction filtration and should be sucked as dry as possible. Further drying may be effected in air.

Place the crystals in a dry, tared bottle and weigh them. Label the bottle with your name and the name of the contents and turn it in to your instructor.

If a suitable melting point bath is available, take a melting point of your sample.

In a future experiment you may determine the molecular weight of your tin(IV) iodide by the freezing point depression method.

SPECIFIC REFERENCES

1. T. Moeller and D. C. Edwards in "Inorganic Syntheses", IV, J. C. Bailar, ed., McGraw-Hill Book Co., New York, N.Y., 1953, p. 119.
2. W. G. Palmer, "Experimental Inorganic Chemistry", Cambridge University Press, Cambridge, 1962, p. 246.
3. F. A. McDermott, "Preparation of Stannic Iodide and Its Solubility in Certain Organic Solvents", *J. Amer. Chem. Soc.*, 33, 1963 (1911).

EXERCISES

1. A convenient laboratory method for the preparation of anhydrous iron(III) chloride involves the direct reaction of iron with dry chlorine. In an experiment, 30.0 g of iron wire was reacted with chlorine. The $FeCl_3$ sublimed out of the reaction chamber into a flask with a yield of 84.0 g. (a) Calculate the theoretical yield of $FeCl_3$. (b) Calculate the percent yield.

2. Chromium(II) acetate, $Cr(C_2H_3O_2)_2 \cdot H_2O$, is prepared by the reduction of a Chromium(III) solution with zinc and the immediate transfer of this chromium(II) solution into a sodium acetate solution. The red chromium(II) acetate precipitates immediately.

$$2\ CrCl_3\ (aq)\ +\ Zn\ (s) \longrightarrow 2\ CrCl_2\ (aq)\ +\ ZnCl_2\ (aq)$$

$$CrCl_2\ (aq)\ +\ 2\ NaC_2H_3O_2\ (aq) \longrightarrow Cr(C_2H_3O_2)_2 \cdot H_2O\ (s)\ +\ 2\ NaCl\ (aq)$$

In an experiment, 29.9 g of chromium(III) chlordie hexahydrate in 50 ml of 0.4 M H_2SO_4 is reduced with excess zinc and the resulting solution reacted with 84.0 g of sodium acetate in 110 ml of water. The yield of chromium(II) acetate was 18.0 g. (a) Calculate the theoretical yield of $Cr(C_2H_3O_2)_2 \cdot H_2O$. (b) Calculate the percent yield.

(Answers are on p. 246)

REPORT SYNTHESIS OF TIN(IV) IODIDE

1. Record your observations during the course of the reaction and filtration. Be sure to include the appearance of the reaction mixture before and after the reaction and the color of any crystalline material.

2. (a) What color was the condensate dripping back into the reaction flask? _____

   (b) What caused this color? _____

   (c) How did it get there?

3. Actual yield of SnI_4:

   Bottle and SnI_4 _____

   Bottle _____

   SnI_4 _____

4. Melting point range of your sample _____

   The literature value is 144.5°C. How pure is your sample, using the melting point data as a criterion of purity?

5. Calculate the theoretical yield of SnI_4.

6. Calculate the percentage yield of SnI_4 that you obtained.

Name _____ Section _____ Grade _____

A resin is a synthetic organic polymer that is water insoluble and contains the elements C, H, O, and possibly others. The melamine-formaldehyde polymer is a resin that is prepared by condensation polymerization.

Melamine reacts with formaldehyde initially to form trimethylolmelamine.

$$H_2N-C(N)C-NH_2 \;+\; 3\,H_2CO \;\longrightarrow\; HOCH_2NH-C(N)C-NHCH_2OH$$

melamine   formaldehyde   trimethylolmelamine

Trimethylolmelamine then condenses with other —OH or —NH reactive sites to form the polymer, with elimination of H_2O. Early in the polymerization the product is probably linear but as the reaction proceeds, cross-linking (bond formation between polymer chains) occurs. Cross-linking imparts hardness to the completely polymerized product.

A large percentage of the melamine resin produced is used to form molded objects. Probably the best known is melamine dinnerware, made by mixing the resin with cellulose. Used as an adhesive, the resin forms bonds that are resistant to boiling water. The melamine-formaldehyde resin prepared in this experiment will be isolated in two different forms. One portion of a partially polymerized product will be placed in an oven for final polymerization, resulting in a piece of hard, clear resin. The other portion will be isolated as a white powder. For industrial use, polymers are frequently isolated as powders because it is a convenient form for shipping and storage.

Polyvinyl alcohol is an addition polymer but it is prepared by indirect means because the vinyl alcohol monomer is unstable. It is synthesized by the alcoholysis of polyvinyl acetate, itself an addition polymer.

$$[-CH_2-CH(OOCCH_3)-]_n \;+\; n\,CH_3OH \;\longrightarrow\; [-CH_2-CH(OH)-]_n \;+\; n\,CH_3C(=O)-OCH_3$$

polyvinyl acetate   polyvinyl alcohol

Polyvinyl alcohol is soluble in water by virtue of the polar —OH groups that are attached to the hydrocarbon chain. The water solubility is important in some of its major uses, such as a thickener for certain emulsions and suspensions and a packaging film when water solubility is desirable.

When referring to glass, one normally thinks of the material used to make drinking glasses and window panes. These types of glass are made from various combinations of SiO_2, B_2O_3, Al_2O_3, alkali metal oxides and alkaline earth oxides, or salts that decompose to oxides on heating. In general terms, however, a glass can be any amorphous, hard substance, usually transparent. A large number of metal acetate systems containing two or three compounds can be made into glasses. The acetates are melted and *quenched* (cooled very quickly) so that the melt hardens faster than the ions can reorder themselves to allow crystallization. Lithium, lead, and zinc acetates will form glasses by themselves but, by using mixtures of acetates, lower melting points and greater ease of glass formation are obtained. Some of the acetate systems melt at very low temperatures, 150 to 200°C, as compared to borosilicate glasses, which melt above 600°C. Acetates have the disadvantage of being slightly hygroscopic.

PROCEDURE

I Preparation of a Melamine-Formaldehyde Resin

It would be wise to dirty no more glassware than necessary because the melamine-formaldehyde resin is rather difficult to remove. To clean the glassware, fill it with cold water as soon as you are finished with it and allow it to soak. Some of the polymer will flake off and the rest may be removed by vigorous scrubbing with a cleansing powder. If the polymer has cured on the wall of the reaction flask, you may need special cleaning directions from your instructor.

A 125 ml Erlenmeyer flask is used as a reaction flask and the chemicals should be introduced through a powder funnel (a short-stemmed funnel with a large-diameter stem) so the chemicals do not get all over the walls of the flask. A piece of paper rolled into the shape of a cone can be used as a substitute for a powder funnel.

Place the following chemicals into the Erlenmeyer flask:

1. 12.6 g   melamine
2. 9.0 g   paraformaldehyde
3. 15   ml distilled water

Remove the funnel and stir the reactants sufficiently to mix them thoroughly. They will not dissolve until after heating has been started. Make a mark on the flask at the top of the H_2O level as water must be added periodically to replace that which boils off. The flask may be placed on a tripod or clamped to a ring stand, with a wire gauze under the flask in either case. A small watch glass placed on top of the flask will help reduce water evaporation.

The reaction mixture is heated to gentle boiling with a Bunsen burner; shortly after boiling has begun, the reactants should dissolve, giving a clear, colorless solution. Note the time when this occurs. The polymerization will require approximately 30 minutes of very *gentle* boiling from this time. During the course of the polymerization, stir the reaction mixture occasionally with a stirring rod and replace evaporated water by *slowly* adding distilled water up to the mark on your flask.

Heating should be stopped when the reaction mixture has turned milky white (not just cloudy). So far this procedure has been concerned with the science of polymer chemistry but recognition of the point when the reaction should be stopped is part of the "art" of polymer chemistry. If heating is stopped too soon, a low molecular weight polymer will be obtained. On the other hand, if heating is continued too long, a cross-linked polymer will be formed that is insoluble and incapable of further modification. The reaction mixture turns white because the polymer has reached a sufficiently high molecular weight so that it is no longer soluble in water and is precipitating out. At this point the material can still be manipulated and it will require a certain amount of judgment on your part to decide when this point has been reached.

After the heat has been removed from the reaction flask, allow it to cool enough so that you can handle it without burning yourself and pour the contents of the flask into a 50 ml beaker. When this has cooled for a few minutes, you may notice a layer of water forming on top of the polymer mixture. Pour off the water and discard it. While the polymer is still warm, pour about one-half of it into a two-inch diameter aluminum pan,* add 20 drops (about 1 ml) of glycerine and mix them together as thoroughly as possible. The polymer is "cured" by placing the Al pan in an oven at 100°C for 2 to 3 hours. A hard clear piece of melamine-formaldehyde resin should result.

To the remaining one-half of the polymer from the reaction, add 10 ml of *N, N*-dimethylformamide (DMF). AVOID INHALING THE VAPOR OR CONTACT WITH THE SKIN. Work in a hood or keep the beaker covered with a watch glass. Stir the polymer mixture

---

*If no aluminum pans are available, one can be made by taking a 3-in. circle of heavy duty Al foil and forming it around the bottom of a 150 ml beaker.

with a stirring rod to dissolve the polymer in the DMF. Dissolution may be rather slow and, in this case, can be speeded up by placing the beaker in a warm (60 to 80°C) water bath. A magnetic stirrer will speed up dissolution considerably if one is available. While the polymer is dissolving, fill a 600 ml beaker with about 500 ml of cold tap water and set it aside for future use. Also set up a filter flask and Buchner funnel with filter paper for a suction filtration. (See p. 37 on filtration techniques.) When all the polymer has dissolved, pour the solution dropwise into the beaker of water, stirring vigorously throughout the addition. Allow the precipitated polymer to settle and decant one-half to two-thirds of the water, or as much as you can without losing an appreciable amount of polymer. This will make the filtration faster and easier. The remaining polymer should be suction filtered and washed with a small portion of cold water. Suck air through the polymer to dry it as much as possible and then transfer it to a beaker or watch glass to dry further. Place the dried polymer in a sample bottle labeled with your name and the name of the product and turn it in to your instructor.

II Formation of a Polyvinyl Alcohol Film

A. Place 5.0 g of a commercial preparation of polyvinyl alcohol in a 100 or 150 ml beaker and dissolve in 34 ml of distilled water by warming and stirring. Dissolution will be slow and the polymer must be completely or almost completely in solution before continuing. Watch out for polymer or gel particles that have become swollen with the solvent but have not dissolved. These particles appear as colorless, transparent bodies, having nearly the same refractive index as the solution. When all or almost all the particles have dissolved, cool the solution slightly, and add 18 ml of 95% ethanol with stirring. The resulting solution should be filtered by slowly pouring it onto a wad of glass wool placed in a filter funnel.

Prepare two films as follows. Pour (about half for each film) the filtered solution onto two clean, dry 4-in. square glass plates and spread it to a uniform thickness with a "doctor knife" (see Fig. 12-1) prepared by putting three turns of transparent tape on each end of a stirring rod. Allow the films to air-dry undisturbed (about 1 to $1\frac{1}{2}$ hours). The films can be removed from the glass plates by soaking them in a bath of 50% aqueous acetone (FLAMMABLE SOLVENT: Use in a hood and away from open flames) for 5 to 10 minutes and gently pulling them off the plates; dry in air. Clean your glassware with soap and water, preferably hot water, soon after you have finished using it.

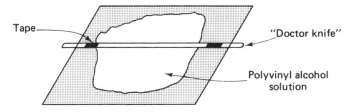

Figure 12-1. Preparation of a film.

III Formation of Glasses from Molten Acetates

B. Choose one of the acetate systems listed below, weigh the proper quantities of chemicals, and mix them together in a 50 ml beaker. Systems (4) and (5) form very strained glasses that crack upon cooling.

(1)   3.2 g $Na(C_2H_3O_2)$ — 8.8 g $Zn(C_2H_3O_2)_2 \cdot 2\,H_2O$

(2)   2.5 g $Na(C_2H_3O_2)$ — 11.4 g $Pb(C_2H_3O_2)_2 \cdot 3\,H_2O$

(3)   3.0 g $K(C_2H_3O_2)$ — 11.4 g $Pb(C_2H_3O_2)_2 \cdot 3\,H_2O$

(4)   3.3 g Na($C_2H_3O_2$) — 8.6 g Mg($C_2H_3O_2$)$_2 \cdot$ 4 H_2O

(5)   3.3 g Na($C_2H_3O_2$) — 9.8 g Mn($C_2H_3O_2$)$_2 \cdot$ 4 H_2O

Have an evaporating dish resting on top of a beaker that is full of cold tap water ready; the bottom of the evaporating dish must be in contact with the water. Heat the beaker containing the acetates gently with a burner (CAUTION: Overheating causes decomposition). The waters of hydration must be driven off first, sometimes this is accompanied by bubbling and melting. Continue heating gently for another 5 minutes or until the mixture is completely molten. If vigorous bubbling develops (after you believe all the water has been evolved) or a brownish color*, stop heating immediately because decomposition is beginning. Quench the molten acetates by quickly pouring the melt into the evaporating dish and allow it to cool to room temperature.

Fiber formation is generally an indication that a system is polymeric. This may be determined by placing a stirring rod in the partially molten acetate remaining in the beaker and slowly pulling the stirring rod from the beaker. Are the acetate fibers weak or strong?

If the acetate glass disks are permitted to be in contact with moist air for several hours, they will become cloudy due to surface crystallization of the metal acetate hydrates. They will remain clear indefinitely if stored in a desiccator.

SPECIFIC REFERENCE

1. R. F. BARTHOLOMEW and S. S. LEWEK, "Glasses Formed from Molten Acetates", *J. Amer. Ceram. Soc.*, **53**, 445 (1970).

GENERAL REFERENCES

F. W. BILLMEYER, Jr., "Textbook of Polymer Science", 2nd ed., Interscience Publishers, New York, N.Y., 1971.

E. L. MCCAFFERY, "Laboratory Preparation for Macromolecular Chemistry", McGraw-Hill Book Co., New York, N.Y., 1970.

W. R. SORENSON and T. W. CAMPBELL, "Preparative Methods of Polymer Chemistry", Interscience Publishers, New York, N.Y., 1961.

C. C. PRICE, "The Geometry of Giant Molecules", *J. Chem. Educ.*, **36**, 160 (1959).

W. P. SLICHTER, "Molecular Characteristics of Rubber-Like Materials", *J. Chem. Educ.*, **36**, 185 (1959).

*The system containing Mn($C_2H_3O_2$)$_2 \cdot$ 4 H_2O normally forms a brown glass.

REPORT POLYMERS

I Melamine-Formaldehyde Resin

1. Time reactants dissolved to give a clear, colorless solution _____

   Time heating was stopped _____

   Total length of heating _____

2. Describe the nature of the polymer reaction mixture at the time heating was stopped.

3. Describe the nature of the final polymers.

   (a) Resin

   (b) Precipitated polymer

4. If you made any changes in the given polymerization procedure, record them below.

5. Sometimes the dried, precipitated polymer becomes sticky or wet after several days or perhaps even after a few hours. Explain what is happening.

6. Record any interesting observations you may have made.

II Polyvinyl Alcohol Film

1. The solvent that was used to prepare the polyvinyl alcohol film was a mixture of ethanol and water. Calculate the *volume percentage* of ethanol in the solvent. Note: 95% ethanol is composed of 95% ethanol and 5% water by volume.

2. Describe the physical properties of your film.

   a gelatinous – clear on glass plate

III Acetate Glass

1. Which acetate system did you study?

   5. 3.3g Na($C_2H_3O_2$) – 9.8 gm Mn($C_2H_3O_2$)$_2 \cdot 4H_2O$

2. Describe the events that occurred during the process of melting the acetates.

   bubbling is first indication, followed by a moltenization of acetate (honey-brown color). Upper surface of acetate is clear. Moltenization is a slow process.

3. Describe the physical properties of the glass that you obtained.

4. Do you think an acetate glass would make a good substitute for silicate glass in window panes? State your reasons.

Name _____ Section _____ Grade _____

Experiment 13
Synthesis of an Amino Acid

Amino acids are a group of biologically active compounds that contain both an amine group and a carboxylic acid group. Most of the biochemically important ones have the amine and the carboxylate groups attached to the same carbon atom and these are called α-*amino acids*.

$$R-\underset{NH_2}{\overset{H}{C}}-COOH \quad \alpha\text{-amino acid}$$

Amino acids are the "building blocks" of which proteins are composed and there are about 20 different amino acids that are most often found in proteins. The amino acids are bound together through amide linkages and the resulting protein is a condensation polymer. The equation below shows a hypothetical reaction between two amino acids.

$$R-\underset{NH_2}{\overset{H}{C}}-\overset{O}{\overset{\|}{C}}-OH + NH_2-\underset{R'}{\overset{H}{C}}-\overset{O}{\overset{\|}{C}}-OH \longrightarrow R-\underset{NH_2}{\overset{H}{C}}-\overset{O}{\overset{\|}{C}}-\underset{H}{N}-\underset{R'}{\overset{H}{C}}-\overset{O}{\overset{\|}{C}}-OH + H_2O$$

amide linkage

In biological systems the condensation of amino acids is highly complex process; in the laboratory, starting materials other than the amino acids would be required to form the amide shown in the equation above. Some of the amino acids are called *essential* amino acids because the human body is not capable of biosynthesizing them and they must be a part of our diet. They are usually consumed in the form of plant and animal proteins that the body hydrolyzes to obtain the amino acids. Some amino acids are not used for protein synthesis but have other important biological functions, especially in metabolism.

The R side chain of the amino acid may be one of a variety of types of functional groups. The simplest amino acid, where R = H, is glycine and the one to be synthesized in this experiment.

$$\text{H} - \underset{\underset{\text{NH}_2}{|}}{\overset{\overset{\text{H}}{|}}{\text{C}}} - \text{COOH} \quad \text{glycine}$$

The R group may be a hydrocarbon side chain, a hydroxide-containing side chain, a sulfur-containing side chain, a carboxylate group, a basic group, or an aromatic side chain, to list some of the most frequently occurring possibilities. All the amino acids have trivial names and you can find these and their structures in biochemistry and organic chemistry textbooks.

Amino acids contain both an acidic group, —COOH, and a basic group, —NH$_2$.

$$-\text{COOH} \rightleftarrows \text{COO}^- + \text{H}^+$$

$$-\text{NH}_2 + \text{H}^+ \rightleftarrows -\text{NH}_3^+$$

In a strongly acidic solution, an amino acid will have the ionic form *A*. In approximately neutral solution, it will exist as the dipolar ion *B*, called a *zwitterion*, which is electrically neutral. In strongly basic solution, the NH$_3^+$ will lose a proton and the amino acid will have the ionic form *C*.

$$\underset{A}{\text{R} - \underset{\underset{\text{NH}_3^+}{|}}{\text{CH}} - \text{COOH}} \qquad \underset{B}{\text{R} - \underset{\underset{\text{NH}_3^+}{|}}{\text{CH}} - \text{COO}^-} \qquad \underset{C}{\text{R} - \underset{\underset{\text{NH}_2}{|}}{\text{CH}} - \text{COO}^-}$$

If one starts with glycine in form *A* and titrates it with standard sodium hydroxide, the titration curve shown in Fig. 13-1 is obtained.

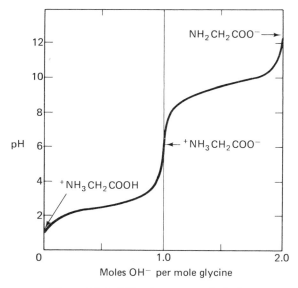

Figure 13-1. Titration curve of glycine.

Two end points are obtained; the first when the proton has been removed from the carboxylic acid group and the second when the proton has been removed from the NH$_3^+$ group. From the curve it can be seen that the dipolar ion, NH$_3^+$CH$_2$COO$^-$, is the predominant

138

species in the pH range of 4 to 9. Therefore, if solid glycine, NH_2CH_2COOH, is dissolved in H_2O, the dipolar ion will be formed.

Amino acids can be synthesized in the laboratory by straightforward chemical reactions. You will synthesize glycine by reacting ammonia with chloroacetic acid.

$$2\,NH_3 + Cl-CH_2-\overset{\overset{\displaystyle O}{\|}}{C}-OH \longrightarrow {}^+NH_3-CH_2-\overset{\overset{\displaystyle O}{\|}}{C}-O^- + NH_4^+ + Cl^-$$

In the preparation of the reaction solution, ammonium carbonate, $(NH_4)_2CO_3$, is dissolved in water with the evolution of gas. Looking at the formula for ammonium carbonate, two possibilities for the identity of the gas should come to mind and two quick tests can be used to establish its identity. Ammonia is a basic gas and can be detected by holding a piece of moist red or neutral litmus in the vapors. It will turn the litmus blue. Carbon dioxide can be confirmed by the formation of a white precipitate when it comes in contact with a calcium hydroxide solution, according to the equation.

$$CO_2\,(g) + Ca(OH)_2\,(aq) \longrightarrow CaCO_3\,(s) + H_2O\,(l)$$

The glycine is isolated from the aqueous solution by vacuum evaporation of some of the water followed by the addition of methanol. Because of its ionic form, glycine is much less soluble in methanol than in water and so it precipitates out of solution. The ammonium chloride that is formed as a by-product of the reaction of ammonia and chloroacetic acid is generally found to contaminate the precipitated glycine, but it can be removed by recrystallization of the glycine, several times if necessary. To test for the presence of NH_4Cl in the crystalline glycine, it is sufficient to test for the chloride ion by reacting a sample with $AgNO_3$. This is a very sensitive test and a white precipitate of AgCl will form even in the presence of trace amounts of chloride ion.

$$Cl^-(aq) + Ag^+(aq) \longrightarrow AgCl\,(s)$$

PROCEDURE

This experiment requires $1\frac{1}{2}$ laboratory periods but there are periods of time when it requires a minimum amount of attention. During these times work can be done on another experiment. Part A requires about 2 hours and should be done in one laboratory period. Part B should be done in the following laboratory period and requires the full period.

A Preparation of the Reaction Solution

Place 35 g of powdered* ammonium carbonate in a 200 or 250 ml Erlenmeyer flask and add 30 ml of distilled water. Clamp the flask to a ring stand and place it in a warm water bath. A large beaker or other large container can be used to make the water bath. Maintain the bath temperature at 50 to 60°C (no higher!) by heating gently with a burner; stir the ammonium carbonate occasionally to speed up dissolution. You will observe that considerable gas evolution accompanies the dissolution process and although it isn't necessary to the experiment that the gas be identified, let us satisfy any possible scientific curiosity by performing a couple of simple tests.

1. Hold a moist piece of red or neutral litmus paper above the mouth of the flask and observe the results.

*Any lumps present will retard dissolution. Be sure to break up all lumps, however small, by grinding with a mortar and pestle if necessary.

2. Fill a medicine dropper with $Ca(OH)_2$ solution and squeeze the bulb just enough to form a drop on the tip. Carefully, so the drop does not fall into the flask, hold the dropper above the mouth of the flask for 1 or 2 minutes.

Record the results of the tests on the Report sheet. What gas(es) do you believe is (are) being liberated?

When all the $(NH_4)_2CO_3$ has dissolved, remove the flask from the warm water bath. Add 50 ml of 15 M aqueous NH_3 and stir the solution. Prepare a chloroacetic acid, $ClCH_2COOH$, solution in a small beaker by dissolving 10 g of chloroacetic acid (CAUTION: A corrosive chemical, avoid contact with skin; clean up all spills to protect others) in 15 ml of distilled water. Add the chloroacetic acid solution to the $(NH_4)_2CO_3$ solution *dropwise* while continuously stirring the $(NH_4)_2CO_3$ solution. When the reaction solution has cooled to room temperature, stopper it tightly with a rubber stopper and put it aside until the next laboratory period.

B Isolation of the Glycine

Sometimes when a chemist wants to evaporate a solvent in a hurry or at a low temperature, he will do it in a vacuum. The equipment and particular technique he uses may vary from the very simple to the very sophisticated. You need to remove excess ammonia and a large portion of the water from the reaction mixture in order to isolate the glycine. To do this you will use a set up related to that used for a suction filtration (see Fig. 13-2) to carry out a very simple vacuum distillation.

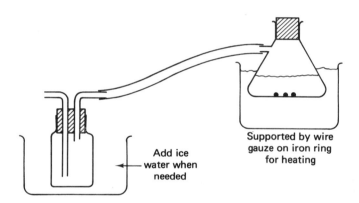

Figure 13-2. Solution in filter flask set up with trap for vacuum evaporation.

Enough solvent should be removed so that the final volume of the solution is about 30 ml. Take a 250 ml filter flask and determine the level to which 30 ml will fill it. Make a mark at this level with some type of water-insoluble marker. Transfer the glycine solution to the filter flask, add some boiling stones, and stopper the flask tightly with a rubber stopper. Connect the filter flask to an aspirator trap (see p. 37) with heavy-walled tubing and connect the trap to the water aspirator. Place the filter flask in a water bath that has been set up so that the water can be heated later. *Slowly* turn on the water aspirator part way; ammonia and dissolved gases will come off rapidly at first. When the gas evolution has subsided, turn on the aspirator all the way. When boiling has again subsided to a small amount, begin heating the water bath, gradually increasing the bath temperature until it is within the range of 50 to 75°C. Boiling should be vigorous, but bumping* is undesirable. If either become so violent that the solution is carried up to the sidearm of the flask, heating must be stopped or the vacuum must be released by removing one of the hose connections. When

*If bumping is violent or continues for more than a few minutes, more boiling stones may be needed.

the volume of solution has reached the 30 ml mark, remove the water bath and release the vacuum slowly.

Decant the solution from the boiling stones into a 200 or 250 ml Erlenmeyer flask. Add 100 ml of methanol and cool the precipitate and the solution *thoroughly* by placing the flask in an ice bath. Suction filter the mixture, collecting the precipitate of glycine in a Buchner funnel. Wash the precipitate with a 20 ml portion of methanol and suck out the excess solvent.

To test for the presence of NH_4Cl in your glycine, place a small amount of the precipitate (about the size of two match heads) in a small test tube and dissolve it in 1 ml of distilled water.

Add 2 drops of 1 M HNO_3 and 2 drops of 0.1 M $AgNO_3$ and mix. A white precipitate indicates the presence of Cl^- and, therefore, the precipitate contains NH_4Cl.

If the chloride ion test was positive, recrystallize the glycine by dissolving it in a minimum amount of water at room temperature. Add methanol (three times the volume of water required to dissolve the glycine), chill thoroughly in an ice bath, and filter and wash the glycine as above. Again test for the presence of NH_4Cl. If the glycine is to be analyzed by the gasometric procedure in Experiment 23, it must be free of NH_4Cl. Dry the glycine in air. Place it in a tared vial, weigh it, and record the weight of glycine obtained.

GENERAL REFERENCES

E. E. CONN and P. K. STRUMPF, "Outlines of Biochemistry", 3rd ed., John Wiley and Sons, Inc., New York, N.Y., 1972, pp. 69-89.

A. L. LEHNINGER, "Biochemistry", Worth Publishers, Inc., New York, N.Y., 1970, Chap. 4.

H. R. MAHLER and E. H. CORDES, "Biological Chemistry", 2nd ed., Harper and Row Publishers, Inc., New York, N.Y., 1971, pp. 43-59.

J. D. ROBERTS and M. CASERIO, "Basic Principles of Organic Chemistry", W.A. Benjamin, Inc., New York, N.Y., 1966, pp. 701-712.

REPORT SYNTHESIS OF AN AMINO ACID

1. Record your observations and conclusions for each of the tests performed on the gas evolved during the dissolution of ammonium carbonate.

    (a) Litmus test

    (b) Ca(OH)$_2$ test

2. Results of the test for NH$_4$Cl impurity in your glycine:

    (a) Original product

    (b) Recrystallized product

3. Yield of pure glycine:

    Weight of bottle and glycine   _____

    Weight of bottle   _____

    Weight of glycine   _____

4. Calculate the theoretical yield of glycine.

5. Calculate the percentage yield of the glycine you obtained.

6. What property of the ammonium chloride causes it to contaminate the glycine?

Name _____ Section _____ Grade _____

Experiment 14
Thermochemical Cycle

The heat flow occurring during a chemical reaction is the heat absorbed (or released) by the system (the chemicals under study). This heat flow is usually measured by having it occur to (or from) the surroundings that are composed of materials of known mass and specific heat. Such heat exchange takes place in an apparatus called a *calorimeter*.

If the reaction is carried out at constant volume (such calorimeters are called *bomb* calorimeters), the resulting heat flow is the change in internal energy, ΔE. In this experiment, the heat flows occur at constant pressure and thus you will be measuring changes in enthalpy, ΔH.

The actual magnitude of the heat flow is dependent on the amount of material used. Thus different experimenters would give different heat flows for the same reaction unless they all used exactly the same amounts of materials. To simplify matters, it is customary to convert the heat flows measured to kilocalories per mole-equation.

There are times, however, when it is impossible (or at best very difficult) to measure heat flows in some chemical or physical changes. In such cases, chemists apply Hess' law directly as shown in Example 1 below or they apply it indirectly by devising a thermochemical cycle. Both internal energy E and enthalpy H are state functions; i.e., they are determined only by the state of the system. Thus if a series of changes are carried out such that the final state is identical with the initial state, then both ΔE and ΔH are zero for the cycle. A classic example of this is the use of a Born-Haber cycle to determine lattice energies (or in some instances, where lattice energies are known, to determine electron affinities). This is illustrated in Example 2 below.

EXAMPLE 1. The reaction of magnesium metal with water to give solid magnesium hydroxide at 25°C is very slow. This heat flow can be calculated, however, if heat flows are known for the reactions of magnesium with hydrochloric acid, of aqueous magnesium chloride with sodium hydroxide, and of hydrochloric acid with sodium hydroxide.

(1) $Mg\ (s)\ +\ 2\ HCl\ (aq) \longrightarrow MgCl_2\ (aq)\ +\ H_2\ (g)$ $\Delta H = -110.4$ kcal

(2) $MgCl_2\ (aq)\ +\ 2\ NaOH\ (aq) \longrightarrow Mg(OH)_2\ (s)\ +\ 2\ NaCl\ (aq)$ $\Delta H = -0.5$ kcal

(3) $HCl\ (aq)\ +\ NaOH\ (aq) \longrightarrow NaCl\ (aq)\ +\ H_2O\ (l)$ $\Delta H = -13.4$ kcal

Taking Reaction (1) plus Reaction (2) minus two times Reaction (3) gives

(4) $Mg\ (s) + 2\ H_2O\ (l) \longrightarrow Mg(OH)_2\ (s) + H_2\ (g)$ $\Delta H = -84.1$ kcal

EXAMPLE 2. Calculate the lattice energy U for zinc chloride, given the following data.

Lattice energy U of $ZnCl_2$ (s):

$Zn^{2+}\ (g) + 2\ Cl^-\ (g) \longrightarrow ZnCl_2\ (s)$

Heat of formation Q of $ZnCl_2$ (s):

$Zn\ (s) + Cl_2\ (g) \longrightarrow ZnCl_2\ (s)$ $Q = -99.5$ kcal

Sublimation energy S of Zn (s):

$Zn\ (s) \longrightarrow Zn\ (g)$ $S = +27.4$ kcal

Dissociation energy D of Cl_2 (g):

$Cl_2\ (g) \longrightarrow 2\ Cl\ (g)$ $D = +58.0$ kcal

Electron affinity A of Cl (g):

$Cl\ (g) + e \longrightarrow Cl^-\ (g)$ $A = -87.3$ kcal

Ionization energies (first plus second) $I_1 + I_2$ of Zn (g):

$Zn\ (g) \longrightarrow Zn^{2+} + 2e$ $I_1 + I_2 = +630.5$ kcal

Let us set up a cycle, starting with $ZnCl_2$ (s): first, convert it to its elements; second, convert the elements to gaseous atoms; third, ionize the gaseous atoms to form gaseous ions; and, finally, react the gaseous ions together to give us back the $ZnCl_2$ (s). For the whole process, ΔH will be zero.

$ZnCl_2\ (s) \xleftarrow{U} Zn^{2+}\ (g) + 2\ Cl^-\ (g)$

$\downarrow -Q \qquad \uparrow I_1 + I_2 \quad \uparrow 2A$

$Zn\ (s) + Cl_2\ (g) \xrightarrow{S + D} Zn\ (g) + 2\ Cl\ (g)$

$\Delta H = 0 = -Q + S + D + I_1 + I_2 + 2A + U$

or

$U = Q - S - D - I_1 - I_2 - 2A$

$= (-99.5 - 27.4 - 58.0 - 216.5 - 414.0 + 174.6)$ kcal

$= -640.8$ kcal

In this experiment, you will measure heat flows for two reactions. Then using these plus two others from the literature, calculate via a cycle the heat flow for the sublimation (with dissociation) of solid ammonium chloride.

146

PROCEDURE

A Determination of the Calorimeter Constant

It would be convenient if the calorimeter (the styrofoam cups) were such a part of the surroundings that it would not absorb or release heat to the material placed in it. But there *is* some heat exchange between the calorimeter and its contents and this must be measured. To determine the magnitude of this heat exchange, a known heat exchange will be carried out in the calorimeter and any deviation from the known value will be attributed to heat absorbed by the calorimeter.

Make a calorimeter by nesting two styrofoam cups. For greater stability, these could be inserted into a 250 or 400 ml beaker. If care is exercised, the thermometer can be used as a stirring rod but always clean it when transferring it from one solution to another. Take temperature readings as accurately as your thermometer will permit. (For thermometers lined in degrees, estimate to the nearest 0.2; if in 0.2°, estimate to 0.05.)

Add 50.0 ml of water to the calorimeter and measure its temperature. In a beaker, heat another 50.0 ml of water to about 70 or 80°C. Measure its temperature and add it *immediately* to the water in the calorimeter. Stir the water and measure its final temperature.

B Heat of Neutralization of HCl (aq) and NH_3 (aq)

Prepare 50 ml of 2 M HCl from the 6 M HCl available and place this in a beaker.

Prepare 50 ml of 2 M NH_3 solution (often called NH_4OH) from the 6 M NH_3 available and place this in a second beaker.

Allow both solutions to reach the same temperature. Then add them to the calorimeter, stir and record the highest temperature reached. If both solutions are not exactly at the same temperature before mixing, you may use an average of the two temperatures.

C Heat of Solution of NH_4Cl (s)   53.5

To the nearest 0.01 g, weigh between 4.4 and 5.2 g of solid ammonium chloride. Place 50.0 ml of water in the calorimeter and record its temperature. Add the NH_4Cl to the water, stir, and record the lowest temperature reached.

D Heat of Sublimation of Ammonium Chloride

Use the data obtained in this experiment and the following data to calculate the heat of sublimation (and subsequent decomposition) of NH_4Cl (s).

$$NH_4Cl\ (s) \longrightarrow NH_3\ (g) + HCl\ (g)$$

$$NH_3\ (g) \longrightarrow NH_3\ (aq) \quad \Delta H = -8.28 \text{ kcal/mole-equation}$$

$$HCl\ (g) \longrightarrow HCl\ (aq) \quad \Delta H = -17.96 \text{ kcal/mole-equation}$$

CALCULATIONS AND ERRORS

It is possible to do calorimetry much more precisely than is done in this experiment. For example, in part B, some of the HCl and NH_4OH solutions will be left in the beakers when these solutions are poured into the styrofoam cups. This error could be avoided by weighing the solutions rather than by measuring volumes. In light of the precision of the thermometer you are using, such errors are not significant. You may assume that water and dilute aqueous solutions have a density of 1.00 g/ml and a specific heat of 1.00 cal/g-deg. You can reasonably expect your value of the enthalpy of sublimation of NH_4Cl to be within ± 10% of the literature value.

EXERCISES Thermochemical Cycle

For water and for dilute solutions, use 1.00 g/ml as the density and 1.00 cal/g-°C as the specific heat.

1. How much heat is released from 50.0 ml of water when it cools from 71.2 to 38.0°C?
2. In a calorimeter, a reaction is carried out that theoretically should release 5424 cal. In an actual experiment, it was found that 5136 cal were absorbed by the water solution during a temperature change of 12.84°C. Calculate the calorimeter constant in calories per degree.
3. When 0.454 g of calcium metal is reacted with 50.0 ml of 1.0 M HCl at 25.9°C, the temperature of the solution rises to 48.2°C. The calorimeter constant has been determined to be 2.68 cal/deg.

   (a) How much heat was released in the reaction as carried out?
   (b) How much calcium reacted (expressed in moles)?
   (c) What is the heat flow per mole of calcium?

4. Set up a thermochemical cycle to calculate the energy of the formation of solid ammonia from gaseous nitrogen and gaseous hydrogen:

   $$\tfrac{1}{2} N_2 (g) + \tfrac{3}{2} H_2 (g) \longrightarrow NH_3 (s)$$

   Use the following information.

   $$\tfrac{1}{2} N_2 (g) + \tfrac{3}{2} H_2 (g) \longrightarrow NH_3 (g) \quad Q_1 = -11.04 \text{ kcal}$$

   $$NH_3 (s) \longrightarrow NH_3 (g) \quad S = +4.13 \text{ kcal}$$

5. In Example 1 in the introduction to this experiment, what would be the heat flow of the reaction of magnesium metal with water if the magnesium hydroxide formed were soluble? (It really is insoluble but we can calculate what the heat flow would be if it were soluble.) Hint: True "spectator" ions, solvated ions that are present both before and after reaction, do not affect the energy of the overall reaction.

(Answers are on p. 247)

REPORT THERMOCHEMICAL CYCLE

A Calorimeter Constant

1. Total volume of water — 100 ml
2. Temperature of ~~one~~ hot water sample — _____
3. Temperature of ~~second~~ cold water sample & calorimeter — _____
4. Final temperature — _____
5. Theoretical final temperature if no heat is absorbed by calorimeter — _____
6. Theoretical − actual temperature, Δt — _____
7. Heat absorbed by the calorimeter (in calories) — _____
8. Temperature change of water sample (No. 4 − No. 2) — _____
9. Calorimeter constant (in calories per degree) — _____

B Neutralization

10. Volume of 2 M HCl — 50 ml
11. Volume of 2 M NH_3 — 50 ml
12. Initial temperature — _____
13. Final temperature — _____
14. Temperature change, Δt — _____
15. Change in enthalpy ΔH (for the amounts used) — _____
16. Amount of HCl reacted (in moles) — _____
17. Change in enthalpy, ΔH, per mole — _____

C Solution

18. Amount of NH_4Cl used (in grams) — _____
19. Amount of NH_4Cl used (in moles) — _____
20. Amount of water used — 50 ml
21. Initial temperature — _____
22. Final temperature — _____
23. Temperature change, Δt — _____

24. Change in enthalpy, ΔH (for the amounts used)

25. Change in enthalpy, ΔH, per mole

D Sublimation

26. Calculate the change in enthalpy, ΔH, in kilocalories per mole for the change

$$NH_4Cl\,(s) \longrightarrow NH_3\,(g) + HCl\,(g)$$

Do not prepare 2M HCL from 6M HCL

$NH_3 + HCL \rightarrow NH_4Cl + \Delta$

The heat balance is the heat of neutralization $\frac{cal}{mole}$
equals Heat to Water to Heat to Calorimeter
equals 100 ml × $\frac{g}{ml}$ × Δt × specific heat

$HCL = 50\,ml \times 2\,\frac{moles}{liter} \times \frac{e}{100\,ml} = .1\,moles$ and .3 mole for 6M

$\Delta + NH_4Cl\,(s) \rightarrow NH_4Cl\,(aq)$ The heat required to dissolve NH_4Cl
= Heat from the water + heat from calorimeter

D Calculate heat of sublimation

$NH_4Cl \rightarrow NH_3\,(g) + HCl\,(g)$ ΔH part E

$NH_4Cl\,(s) \rightarrow NH_4Cl\,(aq)$ ΔH part C

$NH_3\,(aq) + HCl\,(aq) = NH_4Cl\,(s)$ ΔH part B (switch to neg.)

$NH_3\,(g) = NH_3\,(aq)$ ($\Delta H = -8.28\,k\,cal/mole$

17.96

Name _____ Section _____ Grade _____

150

Experiment 12
Polymers

Polymer chemistry is not only a science but also an art. It is a science in that it employs the scientific method to solve a problem but it is an art in that it demands the use of imagination and chemical intuition on the part of the scientist, probably more so than in other areas of chemistry. After you have completed the experiment, you will most likely understand the statement above. You may even feel that magical incantations are necessary in some cases to obtain the desired product.

What is a polymer? A polymer is a large molecule composed of a series of small repeating units. The units are held together by normal chemical bonds and it is the very large size of the polymer molecule (rather than some special type of bonding) that determines its characteristic properties. As the molecular weight of a polymer increases; i.e., as the number of repeating units per molecule increases, the properties gradually change. Generally, the higher molecular weight materials are more stable and have the better properties for commercial applications.

A large variety of polymers exists. There are naturally occurring polymers, such as rubber and cellulose; biological polymers, such as proteins and carbohydrates; inorganic polymers, such as glass and many ordinary inorganic compounds; and synthetic polymers, such as the plastics, resins, and fibers. The synthetic polymers are currently undergoing extensive industrial development because of their many uses in modern technology.

Polymers can have one-dimensional, two-dimensional, or three-dimensional structures. Some examples are given below.

One dimensional polymers (chains):

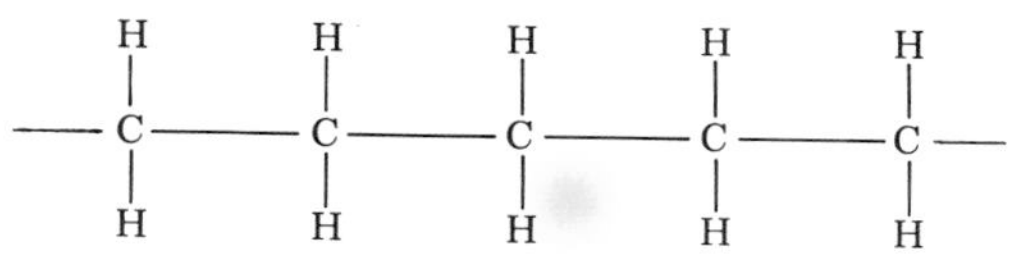

polyethylene

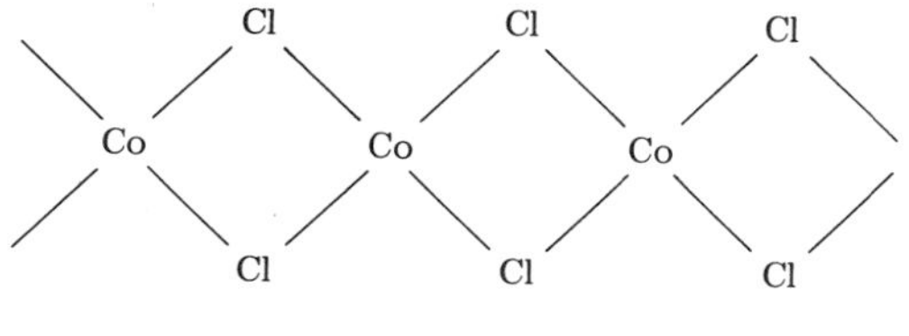

anhydrous $CoCl_2$

Two dimensional polymers (sheets):

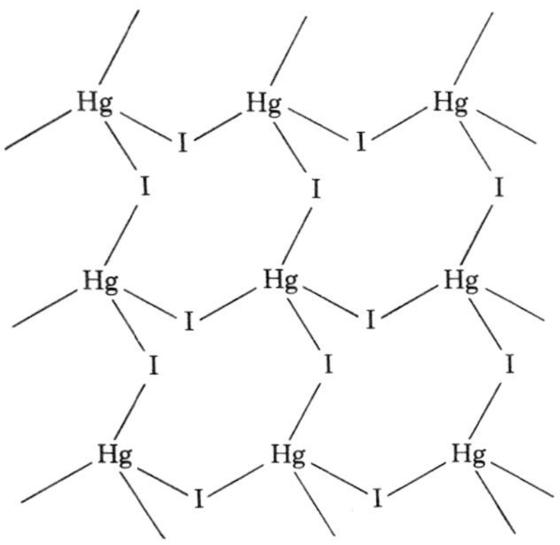

HgI_2

Three dimensional polymers:
glass, melamine resins

There are two general methods of synthesizing polymers; by condensation polymerization and by addition polymerization. Addition polymers are formed by a chain reaction frequently involving either an ion or a free radical. A free radical is a reactive substance that has one unpaired electron. In chain reactions involving free radicals, a small amount of a chemical that readily decomposes to free radicals is used as an initiator to start the reaction. The polymerization reaction is terminated when two free radicals react with each other.

$$R\cdot + H_2C{=}CHCl \rightarrow R{-}CH_2{-}CHCl\cdot$$
(initiator free radical)

$$2\, R{-}CH_2{-}CHCl\cdot + n\, H_2C{=}CHCl \rightarrow R{-}CH_2{-}CH(Cl){-}(CH_2{-}CH(Cl))_n{-}CH(Cl){-}CH_2{-}R$$

polyvinyl chloride

Condensation polymers are formed by the reaction of two molecules to form one larger molecule with the possible elimination of a small molecule such as water or alcohol. The two reacting molecules may be the same or different and the product of the condensation must have sites available for further reaction.

$$H_2N{-}(CH_2)_6{-}NH_2 + HO{-}\overset{O}{\overset{\|}{C}}(CH_2)_6\overset{O}{\overset{\|}{C}}{-}OH \rightarrow H_2N(CH_2)_6NH{-}\overset{O}{\overset{\|}{C}}(CH_2)_6\overset{O}{\overset{\|}{C}}{-}OH + H_2O$$

$$\downarrow$$

$$\left[{-}NH(CH_2)_6NH{-}\overset{O}{\overset{\|}{C}}(CH_2)_6\overset{O}{\overset{\|}{C}}{-} \right]_n + 2n\, H_2O$$

nylon

Procedures for the preparation of a melamine-formaldehyde resin, a polyvinyl alcohol film, and an acetate glass are described in this experiment.

Experiment 15
Determination of a Simple Phase Diagram

In Experiment 3 the relationship among the melting point, melting point range, and the purity of a compound was studied. It was seen that when a second substance was added to a pure compound, the melting point decreased and the melting point range increased. This principle has found wide application in the determination of diverse physical constants. Besides estimating the purity of a compound, it can be used to determine the molecular weight of a substance, the heat of fusion of a substance (in an ideally behaving system), or the phase diagram of a system. In the last three of these, the data are obtained from cooling curves.

A *cooling curve* is a plot of temperature versus time during cooling of a solution or melt from which an accurate determination of the freezing point can be made. Fig. 15-1 shows typical cooling curves for a pure substance and a mixture.

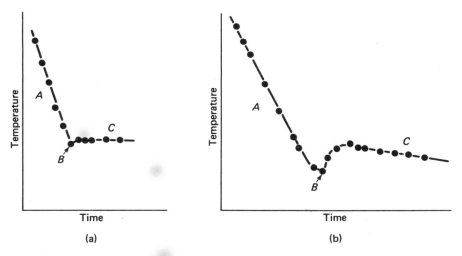

Figure 15-1. Cooling curves. (a) Pure substance, (b) Solution.

The term *freezing point* is synonymous with melting point and is used to denote a temperature of crystallization obtained by *cooling* a system. The upper portion of the curves, A, represents the cooling of the molten substance or solution. The slope of this portion is dependent on the rate of heat loss and the heat capacity of the melt; if the rate of heat loss from the substance(s) to the surroundings is constant, the slope will be constant.

Crystallization is first visually observed at point B but this is not the true freezing point because supercooling has occurred. Supercooling occurs when the melt becomes supersaturated with respect to one of the substances. When precipitation begins, the temperature will increase again due to the heat evolved in the crystallization process. For a pure substance [Fig. 15-1(a)], the temperature will remain constant until all the substance has crystallized, as shown by portion C of the curve, and then it will cool at a rate depending on the rate of heat loss to the surroundings and the heat capacity of the solid.

For a solution of two or more substances [Fig. 15-1(b)], one of the pure substances will begin to crystallize from the solution when it reaches the temperature at which it exceeds its solubility in the solution. Portion C of the curve represents this crystallization. This does not take place at a constant temperature because the composition of the solution (melt) changes as the pure substance precipitates from it.

Because of supercooling it is necessary to use a graphical means to obtain an accurate value for the freezing point. This is done by drawing straight lines through portions A and C of the cooling curve and extending the lines until they intersect, as illustrated in Fig. 15-2.

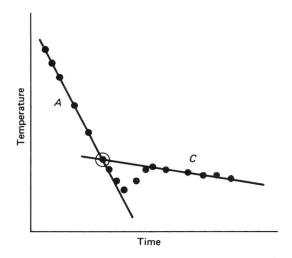

Figure 15-2. Cooling curve showing how to determine the freezing point.

The effect of supercooling is not observed or is observed as a small effect in a pure substance in contrast to the very large effect sometimes noted in solutions. Supercooling can be reduced experimentally by cooling at a very slow rate and by efficient stirring of the melt.

When the freezing points are plotted versus the mole fraction of one of the components of the system, a plot called a *phase diagram* is obtained. A system containing two substances that do not react has a diagram similar to that of the benzoic acid-acetamide system shown in Fig. 15-3. The minimum is called the *eutectic point.* To the left of the eutectic point, the curve expresses the freezing point lowering of pure benzoic acid by the addition of acetamide and, to the right of the eutectic point, the freezing point lowering of pure acetamide by the addition of benzoic acid. The cooling curve for a mixture having the composition at the eutectic point is the same as that for a pure substance except that the eutectic mixture exhibits a larger supercooling effect. In some cases the two substances will react to form a new compound. A typical phase diagram for this kind of system is shown in Fig. 15-4. The maximum occurs at the composition of the new compound and the temperature at the maximum is the freezing point of the pure compound. The minima on either side of the maximum are the eutectic points. There are times when it is inconvenient or impossible to isolate a compound. In these cases it may be advantageous to use cooling curves to determine its composition, freezing point, and range of stability. Phase diagrams can be very complex but provide a wealth of information.

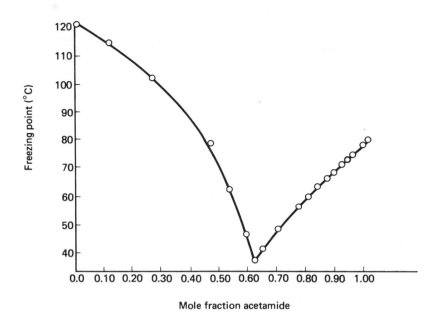

Figure 15-3. Phase diagram for benzoic acid-acetamide system. Original data published by R. KREMANN, O. MAUERMANN, and V. OSWALD, *Monatsh. Chem.*, 43, 335 (1922).

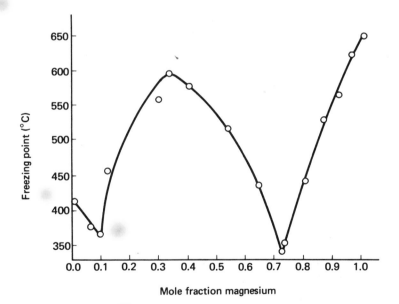

Figure 15-4. Phase diagram for Zn-Mg system. Redrawn from *International Critical Tables of Numerical Data: Physics, Chemistry, and Technology*, National Academy of Sciences, National Research Council Washington, D.C., II, 437 (1923-1928), original data published by G. GRUBE, *Z. Anorg. Chemie*, 49, 72 (1906).

The abscissa of a phase diagram is usually expressed in units of mole fraction or mole percent, probably because one can tell at a glance the molar composition of any point on the diagram.

It is defined as

$$\text{Mole fraction } A = \frac{\text{moles of } A}{\text{moles of } A + \text{moles of } B + \text{moles of } C + \cdots}$$

$$= \frac{\text{moles of } A}{\text{total moles of all substances}}$$

It is designated by the symbol X_A, the subscript referring to the particular component. Mole fraction is related to mole percent by the expression

$$\text{Mole percent} = \text{mole fraction} \times 100$$

GENERAL REFERENCES

D. H. ANDREWS, "Introductory Physical Chemistry", McGraw-Hill Book Co., New York, N.Y., 1970, pp. 214-220.

F. DANIELS and R. A. ALBERTY, "Physical Chemistry", 3rd ed., John Wiley & Sons, Inc., New York, N.Y., 1966, pp. 156-163.

Other introductory physical chemistry texts also contain discussions of this topic.

PROCEDURE

For this experiment you will work in pairs. You and your partner will choose or be assigned one of the systems listed below. Each of you will determine the freezing point at five compositions and then pool your data so that each of you can plot the complete phase diagram. You will both do one composition in common to serve as a check on the consistency of the two sets of data. You will need to work efficiently in order to complete the five freezing point determinations in one laboratory period.

The apparatus consists of a large test tube (175 × 22 mm or 200 × 25 mm) that is placed inside an insulating container. This container may be a larger test tube, an Erlenmeyer flask, a filter flask, a gas wash bottle, or a narrow-mouthed bottle, to name a few of the possibilities. If the test tube rests on the bottom of the insulating container, place a piece of glass wool under the test tube. The purpose of the insulating container is to reduce the cooling rate. A 0° to 150°C (or higher) thermometer and a watch or clock with a second hand are needed.

Most of the chemicals used in this experiment are toxic to a greater or lesser extent so

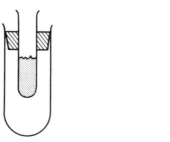 

Figure 15-5. Suggested apparatus setups.

precautions should be taken to avoid contact with the skin or inhalation of vapors from the melt. In particular, resorcinol is a skin irritant and 4-nitrophenol powder is dangerous if inhaled.

Weigh exactly to two decimal places the amount of chemical for Composition I of your system. (In all cases this is one of the pure components of the system.) Carefully transfer it to the test tube and tap the test tube gently on the bench top to knock down any chemical that is adhering to the wall of the tube. The test tube is heated in a *small* flame until the solid is just melted. Swirl the contents *gently* while heating or stir with your thermometer. Overheating should be avoided because (1) the compounds will begin to vaporize or sublime onto the upper part of the test tube and (2) a longer time will be required for the melt to cool down to the freezing point. All the pure compounds listed below melt between 80 and 135°C. It is desirable to have the melt roughly 20°C above its freezing point at the beginning of the time-temperature measurements. Place the test tube containing the melt in the insulating container. Stir nearly continuously and record the time and temperature of the melt each 30 seconds. The measurements may be stopped about 5 minutes after crystallization has been observed.

The compositions listed in Table 15-1 are the total amounts of chemicals that the test tube should *contain* for that freezing point determination. For Compositions II, III, IV, and V, add successive amounts of the second component of your system (it may be component A or B, depending on whether you are Partner No. 1 or No. 2) to achieve the quantities of chemicals listed for each composition. Note that you will not have to add any more of the chemical that you weighed out for Composition I. For each of the compositions, melt and mix the chemicals and determine the freezing point in the same manner as you did for Composition I. At the end of the experiment the chemicals in the test tube should be melted and poured into a special waste container that will be provided.

CALCULATIONS

Plot temperature versus time for each composition and determine the freezing point using the method described on p. 152. Calculate the mole fraction of *component A* for each composition. Obtain from your partner the values for the freezing points and mole fraction of A of his compositions, combine this with your data, and plot freezing point versus mole fraction of A to give the phase diagram. Since Composition V is identical for both partners, your data and your partner's should agree for this point.

EXERCISES

1. How many moles of A are there in 5.00 g of A if A has a molecular weight of 80.00?
2. 10.00 g of B is added to 10.00 g of C. If B has a molecular weight of 50.00 and C has a molecular weight of 200.0, what is the mole fraction of B in the mixture?
3. Referring to Fig. 15-4, what is the mole fraction of Mg in the compound formed in the Zn-Mg system? What is the formula of this compound?
4. Determine the freezing point of a solution for which the following data were obtained.

Time (min)	Temperature (°C)	Time (min)	Temperature (°C)
1.0	115.5	5.0	101.0
1.5	112.5	6.0	102.0
2.0	110.0	6.5	101.0
2.5	107.5	7.0	101.0
3.0	105.5	8.0	100.5
3.5	102.5	9.0	99.5
4.0	100.0	10.0	99.0
4.5	98.0	11.0	98.5

(Answers are on p. 248)

TABLE 15-1

Compositions for Phase Diagrams

System 1

A: Acetamide, $CH_3\overset{O}{\overset{\|}{C}}-NH_2$, M.W. = 59.07 g/mole

B: 4-Nitrophenol, $HO-\langle\bigcirc\rangle-NO_2$, M.W. = 139.11 g/mole

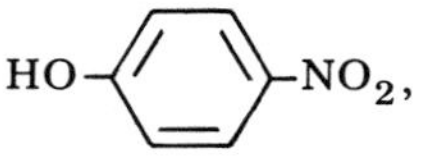

Partner No. 1	Composition	I	II	III	IV	V
A: Acetamide, g of		5.91	5.91	5.91	5.91	5.91
B: 4-Nitrophenol, g of		0.00	1.54	2.64	5.95	13.91
Partner No. 2						
A: Acetamide, g of		0.00	1.48	2.30	3.94	5.91
B: 4-Nitrophenol, g of		13.91	13.91	13.91	13.91	13.91

System 2

A: Benzamide, $\langle\bigcirc\rangle-\overset{O}{\overset{\|}{C}}-NH_2$, M.W. = 121.14 g/mole

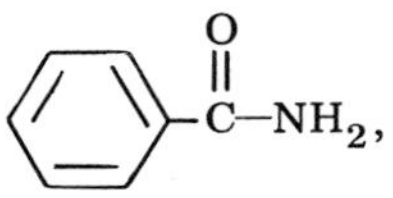

B: Resorcinol, $\langle\bigcirc\rangle-OH$ (HO), M.W. = 110.11 g/mole

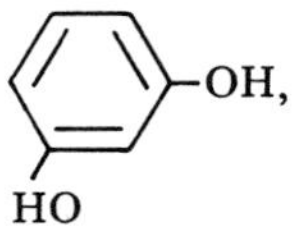

Partner No. 1	Composition	I	II	III	IV	V
A: Benzamide, g of		12.11	12.11	12.11	12.11	12.11
B: Resorcinol, g of		0.00	1.22	4.71	6.20	11.01
Partner No. 2						
A: Benzamide, g of		0.00	1.34	5.18	8.08	12.11
B: Resorcinol, g of		11.01	11.01	11.01	11.01	11.01

System 3

A: Succinimide, (structure), M.W. = 99.09 g/mole

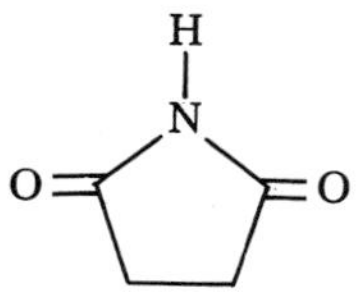

B: Resorcinol, $\langle\bigcirc\rangle-OH$ (HO), M.W. = 110.11 g/mole

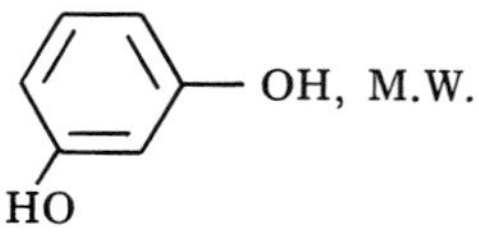

Partner No. 1	Composition	I	II	III	IV	V
A: Succinimide, g of		9.91	9.91	9.91	9.91	9.91
B: Resorcinol, g of		0.00	1.22	2.75	4.71	11.01
Partner No. 2						
A: Succinimide, g of		0.00	1.10	2.48	4.24	9.91
B: Resorcinol, g of		11.01	11.01	11.01	11.01	11.01

REPORT DETERMINATION OF A SIMPLE PHASE DIAGRAM

1. Which system did you investigate? _____
   Partner No. _____

2. Time-temperature data: record time in minutes, temperature in °C.

I		II		III		IV		V	
Time	Temp.	Time	Temp.	Time	Temp.	Time	Temp.	Time	Temp.

3. Plot temperature versus time and determine the freezing point graphically for each composition.

Composition	Freezing point (°C)
I	_____
II	_____
III	_____
IV	_____
V	_____

4. Calculate the mole fraction of *component A* (X_A) of your system for each composition.

   I.

   II.

   III.

   IV.

   V.

5. Plot the freezing point versus mole fraction A and include in this plot the data of your partner.

6. Does compound formation occur in your system?   _____

   If so, what is the freezing point of the compound?   _____

   What is its composition, expressed as X_A?   _____

   What is the stoichiometry of the compound?   _____

Name _____ Section _____ Grade _____

Experiment 16
Heat of Fusion

The measurement of thermodynamic quantities, such as changes in enthalpy, can often be measured directly. An example of this is the calorimetric type of measurement described in Experiment 14: Thermochemical Cycle. There are times, however, when it is impossible, or at best difficult, to do this and it is in such cases that we must resort to indirect methods of measurement. This is not to imply that a theoretical calculation is necessary. Actual measurements can be made and well-established thermodynamic relationships applied to the data to obtain the desired energy relationships.

The equilibrium constant K for a system at equilibrium is dependent on temperature; the relationship between these two quantities is

$$\ln K = -\frac{\Delta H^\circ}{RT} + \frac{\Delta S^\circ}{R} \tag{1}$$

where ΔH° is the standard enthalpy change, ΔS° is the standard entropy change, R is the gas constant, and T is the absolute temperature.

If we make the approximation that ΔH° and ΔS° are constant over the temperature range being studied, for K_1 and K_2 at temperatures T_1 and T_2, respectively, we can write

$$\ln K_1 = -\frac{\Delta H^\circ}{RT_1} + \frac{\Delta S^\circ}{R} \tag{2}$$

$$\ln K_2 = -\frac{\Delta H^\circ}{RT_2} + \frac{\Delta S^\circ}{R} \tag{3}$$

Subtracting Eq. (2) from (3), we obtain

$$\ln \frac{K_2}{K_1} = -\frac{\Delta H^\circ}{R}\left(\frac{1}{T_2} - \frac{1}{T_1}\right) \tag{4}$$

If the equilibrium constant can be measured at two different temperatures (i.e., K_2 at T_2 and K_1 at T_1), Eq. (4) can be used to calculate ΔH°.

In making measurements, it's not a good idea to put all your eggs in one basket and if it is possible to make more measurements and incorporate them all into the calculation, one has more confidence in the result. Equation (1) has the form of a straight line if $\ln K$ is plotted versus $1/T$. From such a plot, $\Delta H°$ can be determined from the slope (since the slope $= -\Delta H°/R$) and $\Delta S°$ from the ordinate intercept (which is $\Delta S°/R$). The greater the number of points in the plot, the greater is the certainty with which the line can be drawn.

Having found a way to determine both $\Delta H°$ and $\Delta S°$, it is now possible to calculate $\Delta G°$, the change in the Gibbs free energy.

$$\Delta G° = \Delta H° - T \Delta S° \qquad (5)$$

$\Delta G°$ can be calculated at any temperature T within the range of applicability of our approximation.

In this experiment you will measure the enthalpy of fusion of naphthalene by the indirect means just described. To do this, we need to have some way of measuring the equilibrium constant for the system

$$\text{naphthalene (solid)} \rightleftarrows \text{naphthalene (liquid)} \qquad (6)$$

at different temperatures. This can be accomplished easily if we study the equilibrium of solid naphthalene with naphthalene in solution:

$$\text{naphthalene (solid)} \rightleftarrows \text{naphthalene (solution)} \qquad (7)$$

This equilibrium may be written in two steps.

$$\text{naphthalene (solid)} \rightleftarrows \text{naphthalene (liquid)} \rightleftarrows \text{naphthalene (solution)} \qquad (8)$$

The enthalpy change of the first step is the heat of fusion of naphthalene, which is the quantity we want to determine. The enthalpy change of the second step will be zero if we choose a compound that will form an ideal solution with naphthalene. To do this, we need a compound that is similar to naphthalene so that interactions between molecules in the liquid state are zero. Such a compound is biphenyl.

naphthalene   biphenyl

Equation (7) is a heterogeneous equilibrium and the equilibrium constant is

$$K = [\text{naphthalene (solution)}] \qquad (9)$$

If mole fraction* X is chosen as the unit of concentration, then

$$K = X \qquad (10)$$

where X is the mole fraction of naphthalene in the solution.

Substituting Eq. (10) into (1) and converting to base 10 logarithms, we obtain

$$\log X = -\frac{\Delta H°}{2.303RT} + \frac{\Delta S°}{2.303R} \qquad (11)$$

*See Experiment 15.

In applying Eq. (11), the mole fraction X of naphthalene is determined from the weights of naphthalene and biphenyl used to prepare the solution and the corresponding temperature T is the freezing point of naphthalene in the solution. At the freezing point, naphthalene (solid) is in equilibrium with naphthalene (solution).

PROCEDURE

For this experiment you may work in pairs if so instructed. You will need to work efficiently in order to complete the five freezing point determinations in one laboratory period. The apparatus consists of a large test tube (175 × 22 mm or 200 × 25 mm) that may be placed in an insulating container as described in Experiment 15. Also needed are a 110°C thermometer and a watch or clock with a second hand.

The chemical used in this experiment are toxic so precautions should be taken to avoid contact with the skin or inhalation of vapors from the melt.

Weigh 10.00 g of naphthalene, carefully transfer it to the test tube, and tap the test tube gently on the bench top to knock down any chemical adhering to the wall of the tube. Heat the test tube over a *small* flame until the solid is just melted. Swirl the contents gently while heating or stir with your thermometer. Overheating should be avoided because (1) the compounds will begin to vaporize or sublime onto the upper part of the test tube and (2) a longer time will be required for the melt to cool to the freezing point. Pure naphthalene melts near 80°C and the solutions all melt below that. It is desirable to have the melt roughly 20°C above its freezing point at the beginning of the time-temperature measurements. Place the test tube in the insulating container or clamp it in place. Stir nearly continuously and record the time and temperature of the melt each 30 seconds. The measurements may be stopped about 5 minutes after crystallization has been observed. This is cooling curve 1 in Table 16-1.

Obtain four more cooling curves with 10.00 g of naphthalene and varying amounts of biphenyl. Table 16-1 indicates the quantities of biphenyl to be added successively to the 10.00 g of naphthalene already in the test tube.

TABLE 16-1

Amounts of Naphthalene and Biphenyl

Cooling Curve No.	Naphthalene To Be Added (g)	Naphthalene, Total (g)	Biphenyl To Be Added (g)	Biphenyl, Total (g)
1	10.00	10.00	0.00	0.00
2	0.00	10.00	2.12	2.12
3	0.00	10.00	1.89	4.01
4	0.00	10.00	4.01	8.02
5	0.00	10.00	4.01	12.03

At the end of the experiment the chemicals in the test tube should be melted and poured into a special waste container that will be provided.

CALCULATIONS

Plot temperature versus time for each composition and determine the freezing point using the method described in the introduction of Experiment 15.

Plot log X (as ordinate) versus $1/T$ (as abscissa) and draw the best straight line through the points. Determine the slope (see p. 7) and the ordinate intercept and, from these, $\Delta H°$ and $\Delta S°$.

EXERCISES

1. How many moles are there in 7.38 g of benzamide (M.W. = 121.14 g/mole)?
2. If 3.00 g of acetamide (M.W. = 59.07 g/mole) is mixed with 10.00 g of benzoic acid (M.W. = 122.12 g/mole), what is the mole fraction of benzoic acid in this mixture?
3. If $X = 0.65$, what is $\log X$?
4. From a plot of $\log X$ of acetamide versus $1/T$ for the acetamide-benzoic acid system, a student determines a slope of $-763°K$. What is $\Delta H°_{fusion}$ of acetamide?
5. That same plot in Exercise 4 gave a $\log X$ intercept of 2.16. What is $\Delta S°$?

(Answers are on p. 249)

REPORT HEAT OF FUSION

1. Cooling curves:

Time (sec)	1 10.00 g of naph. 0.00 g biphenyl	2 10.00 g of naph. 2.12 g biphenyl	3 10.00 g of naph. 4.01 g biphenyl	4 10.00 g of naph. 8.02 g biphenyl	5 10.00 g of naph. 12.03 g biphenyl

2. Calculation of mole fraction of naphthalene:

Curve No.	Naphthalene g	Naphthalene mole	Biphenyl g	Biphenyl mole	Total mole	Mole fraction of naphthalene
1	10.00		0.00			
2	10.00		2.12			
3	10.00		4.01			
4	10.00		8.02			
5	10.00		12.03			

3. Calculations for $\log X$ versus $1/T$ plot:

Curve No.	X	$\log X$	$t\,(°C)$	$T\,(°K)$	$(1/T) \times 10^3$
1					
2					
3					
4					
5					

4. Calculation of enthalpy of fusion:

(a) Plot $\log X$ versus $1/T$, setting up your graph so that your $1/T$ data utilize as much of the abscissa available as possible.

(b) Determine the slope of the line (show your calculations).

(c) From the slope, calculate $\Delta H°$ in kilocalories per mole.

5. Calculation of entropy of fusion:

   (a) Replot log X versus $1/T$, setting up your graph so that the abscissa begins at zero.

   (b) Determine the intercept with the log X axis.

   (c) From the intercept, calculate $\Delta S°$ in calories per deg mole

6. Calculation of free energy of fusion:

   For a temperature of 25°C, calculate $\Delta G°$.

Name _____ Section _____ Grade _____

Experiment 17

Molecular Weight Determination by Freezing Point Depression

The freezing point of a solvent is known as a *colligative* property. This property is dependent on the number of particles rather than the nature of the particles. It is possible to use such a property to determine the molecular weight of a solute in a solution. This experiment is another example of how the melting or freezing point of a substance or solution can be used to obtain valuable information. In Experiment 3 it was used to estimate the purity of a compound; in Experiment 15, to determine a phase diagram; in Experiment 16, to determine the heat of fusion.

The addition of a solute to a solvent causes the freezing point of the pure solvent to be lowered. The extent of this lowering is dependent on the quantity of solute added and the nature of the solvent itself. The freezing point lowering is related to the quantity of solute by the expression

$$\Delta T_f = K_f m \tag{1}$$

where ΔT_f is the difference between the freezing point of the pure solvent and the freezing point of the solution, K_f is the freezing point constant for the solvent, and m is the *molality* of the solution. Equation (1) holds for dilute solutions of nonvolatile nonelectrolytes. If the solute is an electrolyte, then it is necessary to take into account the number of ions into which it dissociates in the solution. The freezing point of a 1.0×10^{-2} molal solution of sodium chloride in water theoretically will be lowered two times as much as that of a 1.0×10^{-2} molal solution of sucrose in water because the NaCl dissolves in water forming two particles, Na^+ and Cl^- ions, while the sucrose forms one particle.

Molality is defined as the number of moles of solute n dissolved in 1 kg (1000 g) of solvent.

$$m = \frac{n_{\text{solute}}}{\text{kg}_{\text{solvent}}} = \frac{g_{\text{solute}}}{(\text{M.W.}_{\text{solute}})(\text{kg}_{\text{solvent}})} \tag{2}$$

Substituting Eq. (2) for m in Eq. (1) gives

$$\Delta T_f = \frac{(K_f)(g_{\text{solute}})}{(\text{M.W.}_{\text{solute}})(\text{kg}_{\text{solvent}})} \tag{3}$$

These determinations are most accurate when the solute is a nonelectrolyte and the solutions are very dilute since the interactions between solute particles in very dilute solutions become quite small.

When the solution becomes very dilute, however, the change in the freezing point becomes very small and thus difficult to measure accurately. These problems can be circumvented by measuring an apparent molecular weight at several different concentrations, plotting apparent molecular weight versus concentration, and extrapolating the line to zero concentration at which point the apparent molecular weight should equal the true molecular weight.

Because of supercooling, the temperature at which crystals are first observed during the cooling of a liquid and the temperature of the true freezing point may not be identical. For this reason, cooling curves are used to determine the freezing point accurately. A cooling curve is a plot of temperature versus time during the cooling of a solution or melt. For the method used to determine a freezing point from a cooling curve, see the introduction in Experiment 15.

There are more accurate instrumental methods for the determination of the molecular weight of a compound than this method and if large numbers of determinations were done and greater accuracy were required, one of these other methods would be employed. The freezing point lowering method is simple, however, and requires no expensive or elaborate equipment. It is used by research scientists who may need to determine a molecular weight occasionally.

In this experiment you will determine the molecular weight of a nonionic, nonvolatile solid. It may be an unknown or it may be the SnI_4 prepared in Experiment 11. The experimental value will only be approximate since the solution will have a finite concentration and the thermometer used does not have an expanded scale. Benzene has been chosen as a solvent because it has a melting point such that it can be easily frozen in an ice bath as well as a reasonably large molal freezing point constant, 5.10 deg/molal. Do not be concerned if the freezing point of the benzene is not the same as the literature value or as your neighbors. Each thermometer may be calibrated somewhat differently. The important factor is the temperature *difference* on the same thermometer; absolute values do not matter so long as they are measured carefully and consistently.

PROCEDURE

Carefully pipet 10.00 ml of benzene into a 30 ml test tube. Be certain to use a rubber bulb since *benzene is poisonous.* CAUTION: benzene is flammable.

Determine the cooling curve of pure benzene by immersing the test tube in an ice bath made with ice and water. Take temperature readings every 15 seconds (with a thermometer graduated in 0.1 or 0.2° intervals) as a function of time until the benzene is solid. Use the thermometer as a stirring rod but do it carefully. Take the test tube out of the ice bath and allow the benzene to warm up.

Weight to the nearest milligram a 0.9 to 1.1 g sample of your unknown using a weighing pan or glazed paper and transfer the sample quantitatively to the test tube. [If your instructor directs you to use the tin(IV) iodide, use only about 0.8 g weighed to the nearest milligram.] Be careful not to get any sample on the wall of the test tube. Stir the solution until the solid dissolves, being certain to get all the sample dissolved. Do not weigh much more than the quantity directed or the freezing point might go beyond the thermometer scale or the solubility may be exceeded. Determine the cooling curve for the solution in the same manner as for pure benzene.

EXERCISES

1. What is the weight of 17.45 ml of a liquid that has a density of 1.034 g/ml?
2. What is the molality of a solution prepared by dissolving 1.108 g of naphthalene, $C_{10}H_8$, in 15.00 g of a solvent?

3. What is the molecular weight of a substance, 2.91 g of which lowers the melting point of 50.0 g of camphor from 178.4 to 151.4°C? The freezing point constant for camphor is 37.7°C/molal.
4. What effect will there be on the determination of molecular weight in this experiment if your thermometer reads 0.5°C too high in the range 0 to 100°C?
5. See Exercise 4 of Experiment 15.

(Answers are on p. 250)

REPORT MOLECULAR WEIGHT BY FREEZING POINT DEPRESSION

1. Unknown number _____

2. Data for freezing point of benzene:

Time	Temperature (°C)	Time	Temperature (°C)	Time	Temperature (°C)

3. Weight of container + sample (g) _____

    Weight of container (g) _____

    Weight of sample (g) _____

4. Data for freezing point of benzene (solution):

Time	Temperature (°C)	Time	Temperature (°C)	Time	Temperature (°C)

5. Plot your temperature versus time data and determine

    (a) the freezing point of benzene _____ °C

    (b) the freezing point of the solution _____ °C

    (c) the change in freezing point _____

6. Volume of benzene _____ 10.00 ml _____

    Density of benzene _____ 0.879 g/ml _____ 8.79 g.

    Weight of benzene _____ 8.74 g

7. Calculate the apparent molecular weight of your unknown.

   $\Delta T_F = 4°C$

   $\varphi = \dfrac{5.5(.9)}{x}$

   $m = \dfrac{\Delta T_F}{K_f}$

   $\dfrac{4}{5.5} = .73 \times .0087$

   # moles solute $= .73 \times 10^{-3}$

   $= 6.351 \times 10^{-3}$

   $\dfrac{.9}{6.35 \times 10^{-3}} = 141.7$

8. Tin(IV) iodide: If you determined the molecular weight of the SnI_4 that you previously prepared, compare your experimental value with the theoretical molecular weight. In benzene, $FeCl_3$ forms a dimer, Fe_2Cl_6; does SnI_4 dimerize?

Name _____ Section _____ Grade _____

Experiment 18

An Acid-Base Titration

Quantitative determinations involving acids and bases are best accomplished by means of titration. This is the process of adding a solution (the titrant) from a buret until all the other reactant has reacted. The end point of the titration may be detected by means of a visual indicator as done in this experiment, potentiometrically as in Experiment 19, or by other instrumental methods.

The point in a titration where theoretically equivalent amounts of acid and base have reacted is called the *equivalence point*. The measured value for that point is called the *end point* and, for an accurate determination, the two points should match. The indicator chosen for an acid-base titration is determined by matching the pH range in which the indicator changes color with the pH range around the equivalence point. In this experiment, a weak acid is being titrated with a strong base. For this situation the equivalence point occurs in slightly basic solution and so phenolphthalein, which changes color in the pH range 8.0 to 9.8, is chosen.

If a standard solution of a base is reacted with a sample containing an acid whose formula is known, it is possible to determine the amount of that acid present in the sample (Part II of this experiment). If the formula of the acid is not known, it is still possible to determine something about that acid. If the acid is part of a mixture, it is possible to determine the amount of acid in terms of the number of equivalent weights of acid. If the acid is pure, it is possible to determine the equivalent weight of the acid (Part III of this experiment).

The topic of equivalents, equivalent weight, and normality is often confusing so a few words are offered here in an attempt to clarify this muddy topic. Historically it has been a

TABLE 18-1

Comparison of Definitions

Number of moles = grams per molecular weight	Number of equivalents = grams per equivalent weight
Molecular weight = grams per mole	Equivalent weight = grams per equivalent
Molarity, M, = moles per liter = $\dfrac{g/M.W.}{liter}$	Normality, N, = equivalents per liter = $\dfrac{g/\text{equivalent weight}}{liter}$

useful concept and students who prefer to plug numbers into equations without any understanding have always been staunch supporters of the subject. Since it is still with us, and since it occurs in many books, an understanding of it is necessary.

As seen in Table 18-1, the definitions of number of equivalent weights (or number of equivalents), equivalent weight, and normality parallel the definitions of number of moles, molecular weight, and molarity with only one difference: The equivalent weight of a substance is defined as the weight of that substance that will react with one equivalent weight of another substance; this is not always true with moles. Let us illustrate with phosphoric acid, H_3PO_4, and its reactions with NaOH.

$$H_3PO_4 + NaOH \longrightarrow NaH_2PO_4 + H_2O \quad (1)$$

In Eq. (1), 1 mole of H_3PO_4 (98.00 g) will react with 1 mole of NaOH (40.00 g). This can also be stated that one equivalent weight of H_3PO_4 (98.00 g) will react with one equivalent weight of NaOH (40.00 g). In this case, the molecular weights and equivalent weights are identical since the acid and base react in a 1:1 mole ration and no adjustment in the numbers is necessary to obtain a 1:1 equivalent ratio.

$$H_3PO_4 + 2\,NaOH \longrightarrow Na_2HPO_4 + 2\,H_2O \quad (2)$$

In Eq. (2), 1 mole of H_3PO_4 (98.00 g) will react with 2 moles of NaOH (80.00 g). This is a 1:2 mole ratio. In order to obtain a 1:1 equivalent ratio we can divide the molecular weight of H_3PO_4 by 2 and call this new number (49.00) the equivalent weight of H_3PO_4. Then we can state that one equivalent weight of H_3PO_4 (49.00 g) will react with one equivalent weight of NaOH (40.00 g). Notice that this 1:1 equivalent ratio (49.00:40.00 g) is in agreement with the 1:2 mole ratio (98.00:80.00 g).

From these examples you can see that the equivalent weight of a compound cannot always be predicted in advance without a knowledge of the reaction involved. You should review this subject in your textbook and in Exercise 4.

Now that the muddy water of equivalent weight has been stirred around, we can return to the more important business of titrations. The following is intended as a summary of important relationships pertaining to this experiment and is used in answering Exercises.

In an acid-base reaction where 1 mole of acid reacts with 1 mole of base, at the point in the reaction where equal molar amounts have reacted, the number of moles of base used equals the number of moles of acid used.

$$\text{Moles acid} = \text{moles base} \quad (3)$$

If the acid or base is a solid, we can calculate the number of moles of acid or base.

$$\text{Moles} = \frac{\text{grams of solid}}{\text{molecular weight}} \quad (4)$$

If the acid or base is in solution, we can calculate the number of moles of acid or base.

$$\text{Moles} = \text{molarity} \times \text{volume in liters} \quad (5)$$

In order to analyze an acid, one needs a standard solution of a base. Part I is the standardization of a sodium hydroxide solution using the solid primary standard acid, potassium hydrogen phthalate, $KHC_8H_4O_4$, which has the nickname KHP. The solid is at least 99.97% pure and reacts as a monoprotic acid.

This standard solution of NaOH will be used either to determine the KHP content of a mixture, Part II, or to determine the equivalent weight of an unknown acid, Part III.

GENERAL REFERENCES

I. M. KOLTHOFF, E. B. SANDELL, E. J. MEEHAN, and S. BRUCKENSTEIN, "Quantitative Chemical Analysis", 4th ed., Macmillan Company, New York, N.Y., 1969, Chap. 33, 34.

E. GRUNWALD and L. J. KIRSCHENBAUM, "Quantitive Chemical Analysis", Prentice-Hall, Inc., Englewood Cliffs, N.J., 1972, pp. 66-75, 200-211.

PROCEDURE

Before beginning this experiment, you should reread the section on Liquid Measure paying particular attention to the following.

1. How to fill, read, and use a buret.
2. How to fill and use a pipet.
3. How to make best use of a buret in an experiment by adjusting sample size.
4. How to fill and use a volumetric flask.

I. Preparation of Standard NaOH Solution

From the 8 M NaOH solution available in the lab and distilled water, prepare 500 ml of 0.1 M NaOH solution. Store this in a polyethylene bottle or in a stoppered flask; the plastic container is preferred since basic solutions slowly react with glass. Under no circumstances should a volumetric flask be used. Be sure the solution is well mixed before you use it. At this point you only know that the concentration is approximately 0.1 M. Determine the exact concentration of the solution as described below.

To the nearest mg, weigh a sample of potassium hydrogen phthalate, $KHC_8H_4O_4$ (4.0 to 4.5 g for a 25 ml buret; 8.0 to 9.0 g for a 50 ml buret). Transfer the sample (see p. 24) to a clean 250 ml volumetric flask* and add about 100 to 150 ml of distilled water. Swirl the flask until the KHP dissolves. The dissolution may be rather slow so be patient. When it is all dissolved, dilute to the mark with distilled water and mix the contents by repeated inversions with shaking and swirling of the inverted flask. Hold the stopper in tightly!

Fill a buret with the 0.1 M NaOH solution; make sure there are no air bubbles in the tip of the buret or just above the stopcock. Run base out of the buret until the level is at 0.00 or below. Record the level of the base, estimating the reading to the nearest 0.01 ml.

Pipet a 25.00 ml aliquot of the standard KHP solution into a 200 or 250 ml Erlenmeyer flask. Dilute with about 25 ml of water (use your wash bottle), washing down the sides of the flask in the process. Add 2 drops of phenolphthalein indicator solution.

Slowly run base out of the buret into the KHP solution, swirling the flask and contents. As you approach the end point, the area in the KHP solution where the drop of NaOH falls will turn pink; then the pink color will disappear as the solution becomes mixed. From this point on, add the NaOH dropwise with constant swirling. Occasionally wash down the sides of the flask with water from your wash bottle. The end point is where 1 drop (or less) of NaOH solution causes the solution to become permanently pink throughout. Essential ingredients for a successful titration include care and patience so don't try to hurry. Record the final buret reading, estimating it to the nearest 0.01 ml.

---

*If you are using a 25 ml buret, a 100 ml volumetric flask may be substituted; a 10 ml pipet should be substituted for the 25 ml pipet also.

Repeat the titration two or three more times using a clean flask each time. After the first titration, the others should go more quickly since you now have some idea of how much base is required per aliquot of KHP solution. The base may be added quickly until you are within 2 or 3 ml of the end point; then change to dropwise addition.

SAVE THE NaOH SOLUTION for either Part II or Part III.

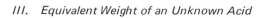

II. Analysis of a Mixture for KHP Content

Before starting this experiment, dry your unknown in a 100°C oven for 30 minutes and then store it in a desiccator until you are ready for it.

To four significant figures weigh a sample of your unknown (0.7 to 0.8 g for a 25 ml buret; 1.4 to 1.6 g for a 50 ml buret) and transfer it to an Erlenmeyer flask. Dissolve it in about 50 ml of water, add 2 drops of phenolphthalein indicator solution, and titrate with your standard NaOH solution.

This first unknown sample may be too large or too small to make the best use of your buret. Repeat the determination two more times using your experience from the first titration to decide on a better sample size.

III. Equivalent Weight of an Unknown Acid

To the nearest 0.1 mg weigh a sample of your unknown acid (0.14 to 0.16 g for a 25 ml buret; 0.28 to 0.32 g for a 50 ml buret) and transfer it to an Erlenmeyer flask. Dissolve it in about 50 ml of water, add 2 drops of phenolphthalein indicator solution, and titrate with your standard NaOH solution. If all the sample does not dissolve, begin the titration anyway, adding base very slowly and swirling continuously. By the time you approach the end point, all the sample will have dissolved (if not, do not continue until it has dissolved).

This first sample may be too large or too small to make the best use of your buret. Repeat the determination two more times using your experience from the first titration to decide on a better sample size.

EXERCISES

1. A 0.4415 g sample of $KHC_8H_4O_4$ required 22.81 ml of NaOH solution for neutralization. What is the concentration of the NaOH solution?
2. A 25.00 ml aliquot of an acetic acid, $HC_2H_3O_2$, solution required an average of 21.28 ml of 0.1163 M NaOH solution. What is the concentration of the acetic acid solution?
3. A 0.8070 g sample of a mixture required 35.75 ml of 0.09939 M NaOH to neutralize the KHP in the sample. Calculate (a) the grams of KHP in the sample and (b) the percent KHP in the sample.
4. A 0.1504 g sample of a pure acid required 18.94 ml of 0.1222 M NaOH solution for neutralization to a phenolphthalein end point.
   (a) What is the equivalent weight of the acid?
   (b) If the acid is a diprotic acid and both protons were neutralized in the titration, what is the molecular weight of the acid?
5. How is it possible to add less than 1 drop of a solution from a buret?

(Answers are on p. 250)

REPORT AN ACID-BASE TITRATION

I. Preparation of Standard NaOH Solution

1. Size of volumetric flask _____ ; size of pipet _____

2. Container + KHP (g) _____

   Container (g) _____

   KHP (g) _____

3. | | 1 | 2 | 3 | 4 |
    |---|---|---|---|---|
    | Final buret reading | ___ | ___ | ___ | ___ |
    | Initial buret reading | ___ | ___ | ___ | ___ |
    | Volume base (ml) | ___ | ___ | ___ | ___ |

   Relative average deviation _____ ppt

4. Calculate the average molarity of the NaOH solution.

27.3406

II. Analysis of a Mixture for KHP Content
(or)
III. Equivalent Weight of an Unknown Acid

5. | | 1 | 2 | 3 | 4 |
    |---|---|---|---|---|
    | Container + unknown (g) | ___ | ___ | ___ | ___ |
    | Container (g) | ___ | ___ | ___ | ___ |
    | Unknown (g) | ___ | ___ | ___ | ___ |

6. | | 1 | 2 | 3 | 4 |
    |---|---|---|---|---|
    | Final buret reading | ___ | ___ | ___ | ___ |
    | Initial buret reading | ___ | ___ | ___ | ___ |
    | Volume base (ml) | ___ | ___ | ___ | ___ |

7. Calculate the percent KHP (II) or the equivalent weight (III).

   Sample 1

Sample 2

Sample 3

Sample 4

8. Summary of results:

    % KHP (II)
    or
    Equivalent weight (III)   Deviation from average

1 _____   _____

2 _____   _____

3 _____   _____

4 _____   _____

Average _____

Relative average deviation _____ ppt

Name _____ Section _____ Grade _____

Experiment 19
Titration Using a ph Meter

It is common practice to use visual indicators to detect the end point during a titration. Such indicators are available for a wide variety of acid-base titrations (see Experiment 18), oxidation-reduction titrations (see Experiment 21), and complexometric titrations (see Experiment 20). Instrumental methods of detection of the end point, however, are generally more accurate, lend themselves to automation, and may also yield additional information if carried out properly. This experiment illustrates how an acid-base titration may be followed using a pH meter and how the end point is determined from the data. It is also possible to estimate the ionization constant of the acid at the same time.

The acid to be titrated is H_3PO_4, which has more than one ionizable hydrogen and thus more than one end point will be detected. Since the third ionization constant is extremely small, only the first two ionizations will be observed. If more sensitive means of measurement (such as conductance) are used, it is possible to detect all three ionizations.

At the first end point, the reaction

$$H_3PO_4 + OH^- \longrightarrow H_2PO_4^- + H_2O$$

will have taken place completely (neglecting any hydrolysis of the dihydrogen phosphate ion). Thus at the point halfway between the beginning of the titration and the first end point, there should be equal concentrations of H_3PO_4 and $H_2PO_4^-$. The first ionization constant for phosphoric acid is

$$K_1 = \frac{[H^+][H_2PO_4^-]}{[H_3PO_4]}$$

At the point where $[H_2PO_4^-] = [H_3PO_4]$, $K_1 = [H^+]$. The pH can be read from the graph at this point and the hydrogen ion concentration calculated from the pH. (*Note:* This only gives an approximate value of K_1 since we again have neglected hydrolysis of H_3PO_4 and $H_2PO_4^-$.)

In a similar manner, the hydrogen ion concentration at the point halfway between the first and second end points is approximately equal to the second ionization constant of H_3PO_4, K_2.

With polyprotic acids, concentrations are sometimes expressed in normality rather than molarity. The laboratory report for this experiment asks you to consider this alternate method of expressing concentration. You will find a discussion of equivalent weights and normality as applied to acids and bases in Experiment 18.

PROCEDURE

Fill a buret with standard NaOH solution (CAUTION: NaOH causes burns) in the prescribed manner. (What is the prescribed manner? If you are unsure, see p. 62.) The NaOH solution is approximately $0.5\ M$; the exact concentration is labeled on the bottle.

Pipet a sample of the H_3PO_4 solution available in the lab into a 250 ml beaker. If you have a 50 ml buret, use a 25 ml pipet; for a 25 ml buret, use a 10 ml pipet. Place a stirring bar into the solution and set up the pH meter with the electrodes in the solution and the buret in position ready to titrate. If necessary, add enough water to the beaker to cover the electrode tips. Be careful to keep the electrodes high enough so that the stirring bar does not bump them.

If you have not used the pH meter before, see p. 46 and request assistance from your instructor; then have him check your setup before you proceed.

Start the magnetic stirrer and read the buret and the pH. Add increments of base from the buret, reading the pH to the nearest 0.01 unit after each increment. (You can keep the magnetic stirrer running throughout the titration.) The increments should be about 0.5 ml at first and in other regions where the pH is changing slowly but 0.05 ml (1 drop) in the vicinity of the end points. Continue until well past the second end point (a pH of about 11.5 to 12).

Plot your data with pH as ordinate and volume of NaOH added as abscissa. Instructions on how to determine the end points of the titration from the graph are on p. 8.

EXERCISES

1. Aliquots of $0.8360\ M$ HCl (25.00 ml) were titrated with an NaOH solution. The average volume of NaOH solution used in three titrations was 38.65 ml. Calculate the concentration of the NaOH solution.
2. An oxalic acid, $H_2C_2O_4$, solution is prepared by dissolving 22.5 g (0.250 mole) in enough water to give 250 ml of solution.
   (a) What is the normality of this solution if it is reacted with base to give $HC_2O_4^-$ as a product?
   (b) What is the normality of this solution if it is reacted with base to give $C_2O_4^{2-}$ as a product?
   (c) What is the normality of this solution if it is reacted with an oxidizing agent such as $KMnO_4$ to give CO_2 as a product?
3. Given below are the data for a pH titration. The titrant is $0.2302\ M$ $HClO_4$ and 10.00 ml of a base is being titrated. Plot the data and determine the volume of titrant needed to reach the end point and the concentration of the base being titrated.

Volume of titrant (ml)	pH	Volume of titrant (ml)	pH	Volume of titrant (ml)	pH
13.03	7.31	13.63	5.89	13.93	2.79
13.13	7.21	13.65	5.56	14.03	2.67
13.23	7.10	13.69	4.40	14.13	2.60
13.33	6.98	13.71	3.80	14.23	2.54
13.43	6.80	13.73	3.48	14.33	2.50
13.55	6.50	13.83	2.98	14.63	2.39

(Answers are on p. 251)

REPORT TITRATION USING A pH METER

1. Molarity of NaOH _____
2. Volume of H_3PO_4 _____
3. Titration data

Buret reading	Volume of NaOH	pH	Buret reading	Volume of NaOH	pH	Buret reading	Volume of NaOH	pH

4. Plot pH (ordinate) versus volume of NaOH (abscissa).

5. From the graph, determine (a) first end point _____ ml; (b) second end point _____ ml; (c) pH at first "halfway point" _____ ; (d) pH at second "halfway point" _____ .

6. (a) Write an equation for the chemical change that occurs to the first end point.

   H_3PO_4 +

   (b) Write an equation for the overall chemical change that occurs to the second end point.

   H_3PO_4 +

7. Calculate the exact concentration (in moles per liter) of the H_3PO_4 solution that you titrated. Should you use the first end point or the second end point or doesn't it matter?

8. (a) Using your data for the first end point, calculate the concentration of the H_3PO_4 solution expressed in normality.

(b) Using your data for the second end point, calculate the concentration of the H_3PO_4 solution expressed in normality.

(c) Why are the answers to (a) and (b) different?

9. From the "halfway points" recorded in No. 5, calculate

(a) the first ionization constant, K_1:

(b) the second ionization constant, K_2:

Compare with literature values: $K_1 = 8 \times 10^{-3}$ and $K_2 = 6 \times 10^{-8}$.

10. Does it make any sense to label a bottle 0.100 N H_3PO_4? Justify your answer.

Name _____ Section _____ Grade _____

Experiment 20
Volumetric Determination of Metals

The quantitative determination of the amount of metal in a sample could be accomplished in a variety of ways. The metal could be precipitated from solution by the formation of an insoluble substance. Gravimetric procedures are often tedious and can be somewhat inaccurate if impurities are trapped in the precipitate. Another method commonly used is electroplating onto a platinum cathode and weighing the metal deposited. This method is accurate but experimentally it may be somewhat difficult. Spectrophotometric methods based on a direct relationship between the intensity of monochromatic light absorbed and the concentration of a complex are widely used. Instrumental methods are available including atomic absorption photometry, which is capable of analyzing for very small amounts of metals in solutions.

Volumetric methods are usually accurate and also have the advantage of being rapid. A recently developed method of quantitatively determining metals is the reaction of metal ions with a chelating agent. The most widely used ligand in such titrations is the anion of ethylenediaminetetraacetic acid. This chelating ligand has been largely applied to the determination of magnesium and calcium in water.

$$H_4EDTA \qquad\qquad EDTA^{4-}$$

EDTA is a good choice of a complexing ligand because it forms very stable complexes with metals and it forms these complexes in a single stoichiometric step. The formation of a 1:1 complex between EDTA and a metal ion results in a sharp change in metal ion concentration at the equivalence point.

The EDTA complexes are usually pale in color and the detection of the end point would be very difficult without an indicator. Indicators are complexing agents that are colored and form colored complexes with metal ions. Such complexes must be weaker (less stable) than those of EDTA so that the metal ion will first react with the EDTA rather than with the indicator.

The divalent metal ion and the EDTA react in the ratio 1:1 according to the equation

$$[M(H_2O)_6]^{2+} + H_2EDTA^{2-} \longrightarrow [M(EDTA)]^{2-} + 2H^+ + 6H_2O$$

It is necessary to maintain a constant pH during the reaction in order to prevent any side reactions involving the water and so a buffer is added to neutralize the protons that are produced during the titration.

Let us first use Cu^{2+} as an example. The solution to be titrated is a copper(II) nitrate solution adjusted to a pH of about 4.9; the color of this solution is blue. Then a small amount of the indicator "Snazoxs" is added. The indicator complexes with some of the copper (not all the copper because only a small amount of Snazoxs is added). The yellow color of the copper-Snazoxs complex plus the blue color of the copper-water complex gives a green (or blue-green) solution. Then a solution of Na_2H_2EDTA is added from a buret. The EDTA complexes the copper forming a blue complex. As the EDTA is added, it first complexes with the copper that was associated with water; then when all that has reacted, the EDTA displaces the Snazoxs from the copper-Snazoxs complex. The blue color of the $[Cu(EDTA)]^{2-}$ complex plus the pink color of the free Snazoxs gives a blue-violet solution. Thus at the end point, the color change is from green to blue-violet (or blue-gray).

If the metal ion being analyzed is Zn^{2+}, there is no color from the zinc ion to effect the color change at the end point of the titration. At the beginning of the titration, before any EDTA has been added, the color of the solution will be that of the zinc-indicator complex. At the end point, the EDTA will displace the indicator from the zinc-indicator complex to give zinc-EDTA complex and free indicator. Since the zinc-EDTA complex is colorless, the final color of the solution will be that of the free indicator.

SPECIFIC REFERENCE

1. G. GUERRIN, M. V. SHELDON, and C. N. REILLEY, "EDTA Titrations Employing 'Snazoxs' as the Indicator", *Chemist-Analyst*, **49**, 36 (1960).

PROCEDURE

I. Analysis of a Copper Oxide Sample

To the nearest 0.1 mg, weigh a sample of your unknown using a weighing pan or a glazed paper. If you are using a 25 ml buret, your sample size should be about 0.10 to 0.11 g; for a 50 ml buret, 0.20 to 0.22 g. (How should you go about weighing out the sample? If you are unsure, see p. 24, Exercise 1.) Transfer the sample to a 250 ml beaker using about 20 to 25 ml of water. (If magnetic stirrers and pH meters are not available, transfer your sample to a 200 ml Erlenmeyer flask rather than a beaker.) Add 10 drops of concentrated HNO_3 and warm over a burner until the sample dissolves. Allow to cool for a few minutes and then dilute with water to about 100 ml. Place a magnetic stir bar into the solution.

Adjust the pH to within the 4.9 to 5.2 range using either pH paper as described below or a pH meter as follows. With stirring, add 1 M NaOH dropwise to the solution and follow the change in pH with the pH meter (turned to READ or pH all during the addition). When the pH has increased to between 2.5 and 3, change to adding 0.1 M NaOH to make the final adjustment. If you add too much base, there is some 0.1 M HNO_3 available to reacidify. Once the pH is adjusted, remove the electrodes, rinse them with distilled water (catch the washings in your sample beaker), and return to your bench. Obtain a magnetic stirrer to stir your solution as you titrate.

Add 2 ml of an acetic acid-sodium acetate buffer.

Add 2 drops of Snazoxs indicator and titrate with the standard EDTA solution. (How accurately should you read the buret? How can you reduce the glare or reflections in the buret while you take a reading? If you are unsure, see p. 32.) The EDTA solution is about 0.01 M; the exact molarity is on the bottle.

The first sample may be too large or too small and you will not know this until you have completed the titration. For your second and third samples, use your results from the first sample to decide on a better sample size.

Report your results as percent copper in the sample.

Adjustment of pH with pH paper. Since the pH does not need to be adjusted very accurately, only within a few tenths of a pH unit, it is possible to use pH paper. It is necessary to transfer some of the solution to the pH paper, however, and if too much of the solution is removed, this will cause an error in the analysis (Would this error make the result too high or too low?). If a small drop is removed on the end of a small stirring rod each time the pH is measured, this error will be small.

After diluting to about 100 ml as described above, check the pH by dipping a small stirring rod into the solution and touch it to a strip of wide-range pH paper. Compare the color formed with the color chart provided. If the pH is less than 4, add several drops of 1 M NaOH, stir, and recheck the pH. Continue this until the pH is close to 4; then use 0.1 M NaOH until the pH is close to 5. When the pH is close to 5, use a narrow range pH paper. If the pH is greater than 5.2, adjust with 0.1 M HNO_3 (1 drop at a time), stirring and checking the pH after each addition.

Once the pH is adjusted, continue with the procedure by adding buffer and indicator as described above.

II. Analysis of a Copper Complex

This procedure is for the analysis of the copper complex, $[Cu(en)_2SO_4]$, which was prepared in Experiment 10. Except for the sample preparation it is the same as the procedure in (I).

To the nearest 0.1 mg, weigh a sample between 0.20 and 0.24 g (if using a 25 ml buret) using a weighing pan or glazed paper. Transfer the sample to a 100 ml beaker using 10 to 15 ml of water. Add 1 ml (20 drops) of concentrated HNO_3. Quantitatively transfer the solution to a clean (does not need to be dry) 100 ml volumetric flask. (For instructions on the quantitative transfer of liquids, see p. 24.) Dilute the solution up to the mark with distilled water, adding the last little bit of water from a dropper. Holding the stopper firmly in the flask, invert 10 to 15 times to mix the solution thoroughly. Pipet a 25.00 ml aliquot of the copper complex solution into a clean 250 ml beaker and dilute with water to about 100 ml. Place a magnetic stir bar into the solution.

Continue with the procedure described in (I), beginning with the paragraph that describes the pH adjustment. Each aliquot should require 17 to 23 ml of 0.01 M EDTA to reach the end point. Analyze at least two, preferably three, aliquots of the solution.

III. Analysis of a Zinc Complex

This procedure is for the zinc analysis of the complex, $Zn(C_5H_7O_2)_2 \cdot H_2O$, which was synthesized in Experiment 10.

To the nearest 0.1 mg, weigh a sample between 0.21 and 0.24 g (if using a 25 ml buret) using a weighing pan or glazed paper. Transfer the sample to a 100 ml beaker using 10 to 15 ml of water. Add 1 ml (20 drops) of concentrated HNO_3. Quantitatively transfer the solution to a clean (does not need to be dry) 100 ml volumetric flask. (For instructions on the quantitative transfer of liquids, see p. 24.) Dilute the solution up to the mark with distilled water, adding the last little bit of water from a dropper. Holding the stopper firmly in the flask, invert 10 to 15 times to mix the solution thoroughly. Pipet a 25.00 ml aliquot of the zinc complex solution into a clean 250 ml beaker and dilute with water to about 100 ml. Place a magnetic stir bar into the solution.

Adjust the pH according to the directions given in (I), except adjust it to within the *pH range 9.0 to 9.5*. Add 2 ml of ammonia-ammonium chloride buffer solution and 2 drops of Eriochrome Black T indicator.

The reaction of the Zn(II) with EDTA in this case is slow, most likely because the acetylacetone is still coordinated to the Zn(II) and must be displaced by the EDTA. The reaction rate may be increased by heating the solution.

To your zinc solution, add about 15 ml of EDTA from your buret. Cover the beaker with a watch glass and heat until it is hot, but DO NOT BOIL. Remove from the heat (How do you handle a hot beaker?) and resume the titration of the hot solution. The reaction still proceeds slowly so titrate very carefully near the end point. Each aliquot should require 17 to 23 ml of 0.01 M EDTA to reach the end point, where the color will change from violet to blue. It may be necessary to use your first aliquot as a trial run to learn to recognize the end point. Analyze at least two, preferably three, aliquots of the zinc solution.

EXERCISES

1. How should you go about weighing the unknown samples for this experiment? Keep in mind that the unknown is to be dissolved in warm nitric acid and thus it is reasonable to assume that it is insoluble in water.
2. (a) Once the sample is dissolved, water is added to make the volume of solution about 100 ml. Does it need to be exactly 100 ml? (b) What if 100 ml is not enough solution to cover the bulb of the glass electrode once it has been adjusted to clear the stirring bar; can more water be added? If so, how much more?
3. Once the pH is adjusted, the electrodes are removed from the solution. Why is it necessary to rinse off the electrodes and to catch the washings in the sample beaker?
4. If 2 drops of Snazoxs is good, why isn't 4 drops better?
5. In a determination of percent cobalt by titration with EDTA, in three separate determinations, the percent cobalt found was 24.35%, 24.05% and 23.90%. Calculate the average percent cobalt, the deviation of each measurement from the average, and the relative average deviation (rad) in parts per thousand (ppt).
6. A solution of Na_2H_2EDTA was standardized by titrating it against 20.00 ml aliquots of a 0.009656 M solution of copper(II) nitrate using the same procedure described in this experiment. The following results were obtained. Aliquot 1 required 18.90 ml, aliquot 2 required 19.00 ml, and aliquot 3 required 18.95 ml of Na_2H_2EDTA solution. Calculate the average molarity of the EDTA solution and the relative deviation in parts per thousand.
7. Some investigators have found an enzyme from spinach that contains 1 mole of copper(II) ion per formula weight. If the formula weight of the enzyme is approximately 5.60×10^5, what size sample would need to be weighed in order to have sufficient copper to react with about 20 ml of 0.010 M Na_2H_2EDTA?

(Answers are on p. 251)

REPORT VOLUMETRIC DETERMINATION OF METALS

I. Weight of samples

	Sample 1	Sample 2	Sample 3
Sample + container (g)	_____	_____	_____
Container (g)	_____	_____	_____
Sample (g)	_____	_____	_____

Unknown number _____

II. or III. Weight of sample

	Part II	Part III
Sample + container (g)	_____	_____
Container (g)	_____	_____
Sample (g)	_____	_____

1.

	Sample 1	Sample 2	Sample 3
Buret reading, final	_____	_____	_____
Buret reading, initial	_____	_____	_____
Volume of EDTA used (ml)	_____	_____	_____

2. Concentration of Na_2H_2 EDTA solution (M) _____

3. Metal(II) ions and EDTA react together in the molar ratio of 1:1. Calculate the percent of metal in your sample.
   (a) Sample 1

   (b) Sample 2

187

(c) Sample 3

4. Calculate your precision in parts per thousand

	% metal	Deviation from average
Sample 1	_____	_____
Sample 2	_____	_____
Sample 3	_____	_____
Average	_____	_____

Relative average deviation: _____ ppt

5. What would be the effect on the accuracy of the determination if the sample size used were about 50 milligrams with all other reagents remaining the same?

6. What would be the effect on the titration if the indicator chosen formed a more stable complex than is formed by EDTA?

7. The amount of copper in a sample prepared by reaction of copper(II) chloride and benzenephosphonic acid was determined by titration with $0.01002\ M\ Na_2H_2$EDTA following a procedure similar to that in this experiment. The first sample weighed 0.03671 g and required 16.80 ml of the EDTA solution; the second sample weighed 0.04100 g and required 18.55 ml of EDTA solution. Calculate the average percent copper in the sample. If the compound was $Cu(C_6H_5PO_3)$, how do the experimental results compare with the theoretical percentages?

Name _____ Section _____ Grade _____

Experiment 21
Vitamin C Determination

During the long voyages of the ancient and not so ancient mariners, scurvy ravaged the crew, frequently killing large numbers of men. Scurvy was also prevalent on land in areas where there was a scarcity of fruits and vegetables. In 1747, Dr. James Lind, a physician in the British navy, discovered that scurvy patients who had oranges and lemons added to their diet recovered very rapidly. At the end of the eighteenth century the British Navy eliminated scurvy in their crews by providing a daily ration of lime juice for each sailor. In 1928, Albert Szent-Gyorgyi isolated pure ascorbic acid for the first time and a few years later proved that this was vitamin C.

Ascorbic acid takes part in a considerable number of physiological functions in humans. It is involved in the metabolism of certain amino acids; it is required in healing processes such as formation of scar tissue and bone deposition; it is required for the conversion of folic acid to folinic acid; and it is necessary for the transport of iron between blood plasma and bone marrow, spleen, and liver. It has been implicated to be linked to several other metabolic functions but these have yet to be clinically or experimentally verified.

Most species can synthesize their own ascorbic acid from glucose. Humans, other primates, guinea pigs, and the Indian fruit-eating bats are exceptions and must obtain a sufficient quantity of this vitamin from their diet. The daily optimum requirement for humans is estimated to be 50 to 100 mg.

Green vegetables and citrus fruits are particularly good sources of vitamin C, with vegetables such as broccoli, brussel sprouts, green pepper, and turnip greens containing more ascorbic acid per 100 g of substance than the citrus fruits. Numerous other fruits and vegetables contain smaller quantities of ascorbic acid. In acidic foods, vitamin C is fairly stable at room temperature or higher temperatures and when these are cooked, canned, frozen, or freeze-dried, they retain 50 to 80% of their original ascorbic acid content. Many vegetables have a copper-containing enzyme, however, that catalyzes the air oxidation of ascorbic acid and these lose ascorbic acid activity readily.

Ascorbic acid is an organic molecule having the structure shown in Eq. (1). It is not a carboxylic acid, as its name would imply, but is a lactone. The protons of the two hydroxyl groups bonded to the ring do have acidic properties because of the ring double bond and the adjacent carbonyl. It is because of this structure that ascorbic acid is very easily oxidized. There are two optical isomers of ascorbic acid, D-ascorbic acid and L-ascorbic acid, but only the L isomer has biological activity.

Several methods have been developed to analyze the amount of ascorbic acid found in fruits and vegetables and the effect of processing upon this quantity. The method[1] that you will use was developed to analyze vitamin tablets or fruit juices whose solutions are colorless or light in color; it utilizes an oxidation-reduction titration. An ascorbic acid solution is titrated with a standard solution of *N*-bromosuccinimide, which quantitatively oxidizes the ascorbic acid to dehydroascorbic acid. The end point is detected by adding a little iodide ion and starch indicator to the ascorbic acid solution. When all the ascorbic acid has been oxidized, the *N*-bromosuccinimide oxidizes the iodide ion to iodine, which complexes with the starch to give a blue color. The equations for the reactions are given below.

Titration reaction:

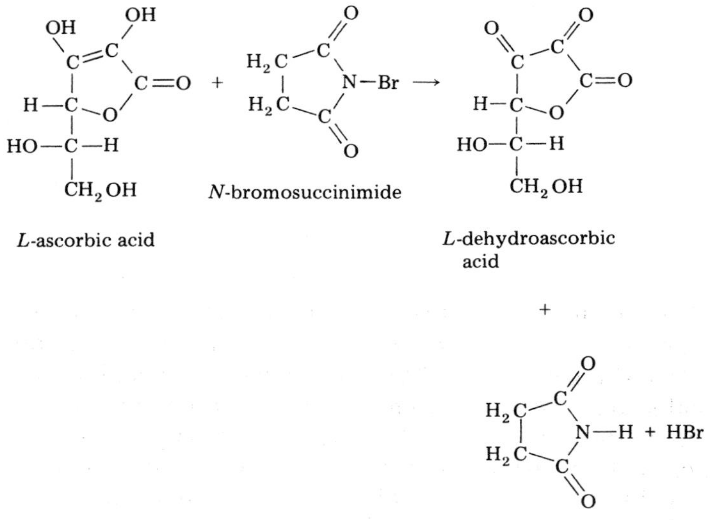

$$L\text{-ascorbic acid} + N\text{-bromosuccinimide} \rightarrow L\text{-dehydroascorbic acid} \quad (1)$$

$$+ \text{ succinimide } + \text{HBr}$$

End point reactions:

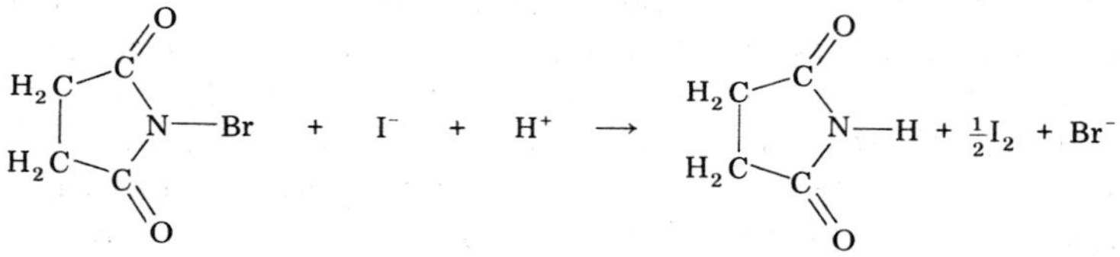

$$N\text{-bromosuccinimide} + I^- + H^+ \rightarrow \text{succinimide} + \tfrac{1}{2}I_2 + Br^- \quad (2)$$

$$I_2 + \text{starch} \rightarrow \text{starch-}I_2 \text{ complex} \quad (3)$$
$$\text{(blue color)}$$

The sites of oxidation in the ascorbic acid molecule are the two $\overset{\diagdown}{\underset{\diagup}{C}}-OH$ groups situated in the ring. Each one of these is oxidized to a $\overset{\diagdown}{\underset{\diagup}{C}}=O$ group with the loss of one electron. Therefore two electrons are lost per ascorbic acid molecule. In the *N*-bromosuccinimide the nitrogen atom is the site of reduction, going from a formal charge of $+1$ in *N*-bromosuccinimide to a formal charge of -1 in succinimide (a two-electron change). One molecule (mole) of *N*-bromosuccinimide is required to oxidize one molecule (mole) of ascorbic acid.

The authors who developed this method of analysis used the *N*-bromosuccinimide as a primary standard. If the chemical used is not pure, this will introduce an error into the analysis. One possible source of error that needs mention is that solutions of *N*-bromosuccinimide are unstable. There is no detectable decomposition after several hours at room temperature or after 48 hours if refrigerated.

Before performing this experiment you should read *Liquid Measure*, p. 29, and *Filtration*, p. 36.

PROCEDURE

I. Analysis of a Vitamin C Tablet

Vitamin C tablets are available with various ascorbic acid contents. The one that you will analyze will be dependent on the size of the buret to be used in the titration. If you will be using a 25 ml buret, a 100 mg vitamin C tablet should be analyzed. If you will be using a 50 ml buret, a 250 mg tablet should be analyzed.

Place a vitamin C tablet in a 100 or 150 ml beaker and add enough distilled H_2O to just cover it. After allowing the tablet to sit in the water to soften for about 5 minutes, carefully pulverize it with a stirring rod. Break up all the lumps into particles as small as you possibly can. The ascorbic acid is water soluble but the filler used to form the tablet may or may not be soluble in water. Dilute the mixture to approximately 50 ml with distilled water and allow it to sit for 10 to 15 minutes, stirring occasionally, to be certain that all the ascorbic acid has dissolved. If any solid particles remain at the end of this time, the solution will need to be filtered. This can be done conveniently by placing a filter funnel above a 100 ml volumetric flask so that the stem of the funnel extends partway into the flask. Fit the funnel with a piece of filter paper and filter the ascorbic acid solution. Rinse the beaker and stirring rod and add the rinse water to the funnel. Finally wash down the filter paper two times with H_2O from your wash bottle.

When the filtration has been completed, remove the funnel and dilute the solution in the volumetric flask to the mark with H_2O. Mix the solution thoroughly by reapeated inversion of the stoppered volumetric flask. Pipet a 25.00 ml aliquot of the solution into an Erlenmeyer flask. Dilute the aliquot to about 75 ml with water, add 5 ml of 4% potassium iodide solution, 2 ml of 3% acetic acid solution, and 3 drops of starch indicator. The solution is ready to be titrated with 0.01 *M N*-bromosuccinimide. (*Note:* The exact concentration is given on the reagent bottle.) Record the buret readings before and after the titration. The end point has been reached when 1 drop of *N*-bromosuccinimide causes the solution to turn a permanent blue color. Repeat the titration on at least one more aliquot of ascorbic acid solution, preferably two more aliquots.

Your results will probably not agree exactly with the value on the label. In the manufacture of tablets, there is an allowable tolerance in the tablet content since mass production methods do not measure the chemical exactly. For vitamin C tablets the allowable limits listed in the U.S. Pharmacopoeia are no less than 95% and no more than 115% of the label value.

II. Analysis of Orange Juice

Just how much ascorbic acid does the juice of an average orange contain? How does the ascorbic acid content of fresh orange juice compare with that of reconstituted frozen orange juice? The purpose of this section is to provide answers to these questions.

The analysis of the juice is not so quantitative a procedure as the analysis of the vitamin C tablet. It is impossible to squeeze *all* the juice from an orange and the frozen juice is not diluted quantitatively. If you are careful, however, you should obtain a reasonable measure of the ascorbic acid content of your juice. You will also be able to check the precision of your titrations; i.e., the agreement between the values for identical aliquots.

CAUTION: Trichloroacetic acid, Cl_3CCOOH, causes burns and may be fatal if taken internally.

A Using Fresh Orange Juice

A juice orange should be rolled vigorously on the bench top prior to squeezing. Obtain a small juice squeezer* and insert it into the stem end of the orange. Squeeze as much of the juice as you can from the orange, collecting it directly in a 100 ml graduated cylinder. This might be a messy procedure but at least you needn't worry about getting squirted in the eye, assuming that you are wearing your safety glasses. Note the quantity of juice obtained. Dilute the juice with an equal volume of 10% trichloroacetic acid, Cl_3CCOOH. The Cl_3CCOOH stabilizes the ascorbic acid and precipitates proteins and other substances, thereby rendering the filtration step a *little* easier.

The pulp of the orange must be removed before the titration, not because it interferes with the actual reaction but because it obscures the end point. In preparation for the filtration, prepare an asbestos mat on a Buchner funnel by filling the funnel one-half full of an asbestos water slurry. Place the funnel on top of a 250 ml filter flask and apply suction to drain the water. Rinse the asbestos mat with distilled water until the filtrate is clear. Clean out the filter flask and replace the funnel. With the suction on, carefully pour the orange juice solution into the Buchner funnel. Rinse the graduated cylinder with a few milliliters of distilled water and add the rinse solution to the Buchner funnel. When all this has been filtered, wash the asbestos mat carefully with two or three 10 to 15 ml portions of distilled water. Because of its pulpy nature, the orange juice may be rather difficult to filter and a slight cloudiness of the filtrate can be tolerated.

Transfer the filtrate to a 250 ml volumetric flask, thoroughly rinsing the filter flask and adding these to the volumetric flask. Dilute to the mark with distilled water and mix the solution by repeated inversions of the stoppered flask. Pipet a 25.00 ml aliquot of the solution into an Erlenmeyer flask and dilute to about 75 ml with water. Add 5 ml of 4% KI solution and 3 drops of starch indicator solution. If you will be using a 25 ml buret for the titration, it is to be filled with 0.002 M N-bromosuccinimide solution. If you will be using a 50 ml buret, it is to be filled with 0.001 M N-bromosuccinimide solution. Note that these N-bromosuccinimide concentrations are different than that used in the analysis of the vitamin C tablet. Titrate the orange juice solution with the appropriate N-bromosuccinimide solution, recording the buret readings before and after the titration and the exact concentration of N-bromosuccinimide used. At the end point the color changes from a yellowish-green to a purple color. Repeat the analysis on at least one more aliquot, preferably two more, of juice solution.

B Using Frozen Orange Juice

In a graduated cylinder, obtain 100 ml of reconstituted frozen orange juice that has been prepared according to the directions on the container. Dilute the juice with 75 ml of 10% trichloroacetic acid. The function of the trichloroacetic acid is to stabilize the ascorbic acid and to improve the filtration process.

With one exception the remainder of the analytical procedure is the same as that given in Part A for fresh juice, beginning with the second paragraph. The reconstituted frozen orange juice is so much pulpier than the fresh juice that two large spatulafuls of wet asbestos should be added to the juice solution and the mixture stirred before filtering it through the asbestos mat.

SPECIFIC REFERENCE

1. M. Z. Barakat, M. F. A. El-Wahab, and M. M. El-Sadr, *Anal. Chem.*, 27, 536 (1955).

---

*Alternatively, a hole 1.5 to 2.0 cm in diameter can be cut in the stem end of the orange. After removal of the center skin and small piece of orange, the juice can be squeezed out by hand via this hole.

GENERAL REFERENCES

C. G. KING, *Nutr. Rev.* 26, 33 (1968).

J. MARKS, "The Vitamins in Health and Disease; A Modern Reappraisal", Little, Brown and Company, Boston, Mass., 1968, pp. 129-136.

L. PAULING, "Vitamin C and the Common Cold", W. H. Freeman and Company, San Francisco, Calif., 1970.

A. K. SIM, "Ascorbic Acid—A Survey, Past and Present", *Chem. Ind. (London)*, 160 (1972).

EXERCISES

1. How many milligrams of ascorbic acid are there in 2.139×10^{-5} mole of ascorbic acid?
2. 100 ml of orange juice is filtered and transferred to a 250 ml volumetric flask. After proper dilution, aliquots are taken using a 25 ml pipet. It is found that each aliquot contains an average of 16.01 mg of ascorbic acid. Calculate the amount of ascorbic acid per milliliter of juice.
3. 10.0 ml of orange juice was found to contain 6.17 mg of vitamin C. How much vitamin C does a 4.0 fluid ounce serving of this juice contain?
4. A vitamin C tablet required 22.72 ml of 0.01274 *M* *N*-bromosuccinimide. How much ascorbic acid did the tablet contain?
5. In a triplicate analysis, a student obtained 7.208 mg, 7.167 mg and 7.178 mg of ascorbic acid per aliquot. Calculate the average amount of ascorbic acid per aliquot and the relative average deviation in parts per thousand.

(Answers are on p. 252)

REPORT VITAMIN C DETERMINATION

I. Analysis of a Vitamin C Tablet

1. Ascorbic acid content as reported by the manufacturer _____

	Sample 1	Sample 2	Sample 3
2. Buret reading, final	_____	_____	_____
3. Buret reading, initial	_____	_____	_____
4. Volume of N-bromosuccinimide	_____	_____	_____

5. Molarity of N-bromosuccinimide _____

6. Calculate the number of milligrams of ascorbic acid in your samples. The molecular weight of ascorbic acid is 176.13.

   (a) Sample 1

   (b) Sample 2

   (c) Sample 3

7. Using the average value of the number of milligrams of ascorbic acid per sample, calculate the amount of ascorbic acid in your vitamin C tablet.

8. Calculate the percent relative error of your value from the manufacturer's value for the tablet.

II. Analysis of Orange Juice   _____

  1. Number of milliliters of juice used: _____

                              Fresh Juice   _____

                              Reconstituted juice   _____

                              Sample 1   Sample 2   Sample 3

  2. Buret reading, final   _____   _____   _____

  3. Buret reading, initial   _____   _____   _____

  4. Volume of N-bromosuccinimide _____ _____   _____

  5. Molarity of N-bromosuccinimide _____

  6. Calculate the number of milligrams of ascorbic acid in your samples. The molecular weight of ascorbic acid is 176.13.

   (a) Sample 1

   (b) Sample 2

   (c) Sample 3

  7. Calculate the number of milligrams of ascorbic acid per milliliter of *juice*.

   (a) Fresh orange juice

   (b) Reconstituted juice

8. Calculate the number of milligrams of ascorbic acid you would ingest by drinking a 4 fluid ounce serving (typical breakfast-size serving) of juice. 1 fluid ounce = 29.57 ml

   (a) Fresh orange juice

   (b) Reconstituted juice

9. Calculate the number of milligrams of ascorbic acid you found in the juice of an orange.

Name _____ Section _____ Grade _____

Experiment 22
A Gravimetric Analysis

The subject of quantitative analysis, the determination of the amount of a given chemical in a sample, is encountered by most scientists and many engineers. When one needs to do a great many analyses, all of which are the same, an instrumental method is developed because such a method is more rapid. At times when it becomes necessary to do only a few quantitative analyses and it is not feasible to use an instrumental method, a volumetric method is the next best choice (see Experiment 20 for an example) and if that doesn't work, one can usually fall back on a gravimetric method.

Gravimetric analysis is considered here for several reasons. One is that historically such methods were responsible for the development and measurement of atomic weights and contributed to the development of our understanding of chemical reactions on an atomic and molecular basis. Another is that such methods or related techniques are a necessary part of sample preparation for other methods of analysis. Thus, it is necessary to learn how to transfer solid samples quantitatively, how to purify samples quantitatively, and how to handle the calculations involved if one is to be any sort of a laboratory scientist.

The basis for gravimetric analysis is the conversion of that portion of a sample in which we are interested into a solid form that can be weighed. In order for this to succeed, that solid needs to be pure and we need to know exactly what it is. Let us consider a method of analysis for calcium ion as an example. It is possible to quantitatively precipitate all (as far as our weight measurements are concerned) the calcium from a solution as a hydrated calcium oxalate, $CaC_2O_4 \cdot H_2O$. It would seem that all we would need to do is dry and weigh this precipitate. The composition of this compound is not as fixed as the formula implies, however, because the amount of water in the compound varies. Thus we really don't know *exactly* what we have. So the method calls for heating this precipitate to around 500°C where it is converted to $CaCO_3$, which meets all our requirements as a stable, pure, well-defined compound. We have to be careful not to heat the $CaCO_3$ too hot because at slightly higher temperatures it loses CO_2 and becomes CaO. Oh well, you might say. Why not just go ahead and convert the whole thing to CaO? That would be fine except that CaO is very hygroscopic (picks up water from the air) and therefore is very difficult to keep anhydrous while being weighed.

Another way in which a product can be unsuitable for our use is if it is impure. Impurities may arise from the presence of other ions that react in a similar manner or by adsorption onto the surface or occlusion within the solid.

The purpose of this experiment is not to learn how to overcome all these difficulties but to learn the techniques involved in handling and transferring solids. Indeed, for a very large number of analyses, procedures have already been established for separations and purifications. Thus this experiment is a straight forward gravimetric analysis[1] that may be used for either a water-soluble magnesium-containing compound or the manganese(II) coordination compound prepared in Experiment 10. The compound to be precipitated is $MNH_4PO_4 \cdot H_2O$ where M = Mg^{2+} or Mn^{2+}.

It is possible to analyze for either Mg^{2+} or Mn^{2+} by the same procedure because both ions often behave chemically in a very similar manner (except for oxidation-reduction reactions). The precipitates formed in this experiment both have about the same solubilities; the hydroxides also are similar.

TABLE 22-1
Solubility Product Constants

	K_{sp}
$MgNH_4PO_4$	2.5×10^{-12}
$MnNH_4PO_4$	1.0×10^{-12}
$Mg(OH)_2$	9.0×10^{-12}
$Mn(OH)_2$	4.6×10^{-14}

In the study of biological reactions that require Mg^{2+}, it is often found that Mn^{2+} can be substituted for the Mg^{2+} and the biological activity still be maintained. The reason for such similarity in chemical behavior can be attributed to the fact that both ions have the same charge (+2) and almost the same size (radius of Mg^{2+} = 0.87 Å; radius of Mn^{2+} = 0.93 Å).[2]

All phosphates are soluble in acids including the metal ammonium phosphates formed in this experiment. Thus it is necessary to carry out the precipitation in neutral or slightly basic solution.

Precipitates that form immediately upon mixing reagents usually are composed of very fine particles. These particles can be very difficult to filter but can be converted to larger particles by a process known as *digestion*. If a precipitate is allowed to stand in contact with its solution, the smaller particles (which are more soluble) dissolve and reprecipitate out onto the larger particles. This process of digestion can be speeded up by heating.

If you are doing a magnesium analysis, the sample supplied to you is water soluble and consists of magnesium sulfate mixed with some inert material. Your results will give you the total magnesium ion content of the sample.

If you analyze the coordination compound $Mn_3(HEDTA)_2 \cdot 10H_2O$, your results may be somewhat unexpected until you learn something about the compound. Experiment 10 emphasized the structure of two out of the three manganese centers as being seven coordinate $[Mn(HEDTA)(H_2O)]^-$ held together by hydrogen bonds. The third Mn(II) in the solid is six coordinate, four positions being occupied by water molecules and the other two by oxygens from the EDTA's lying above and below this manganese. These EDTA's are "wrapped around" the seven coordinate manganese but each carboxyl group uses only one of its two oxygens to coordinate and thus there is an additional oxygen available for coordination to this third manganese. These bonds are weak and are broken when the compound dissolves in water. The hydrogen bonds holding the seven coordinate centers are also broken and the result is the formation of $[Mn(H_2O)_6]^{2+}$ (from the six coordinate center) and $[Mn(HEDTA)(H_2O)]^-$ (from the seven coordinate centers).[3] When $(NH_4)_2HPO_4$ is added to a solution containing these ions, the $[Mn(H_2O)_6]^{2+}$ reacts and forms a precipitate.

$$[Mn(H_2O)_6]^{2+} + NH_4^+ + PO_4^{3-} \longrightarrow MnNH_4PO_4 \cdot H_2O\ (s) + 5\ H_2O$$

But the $[Mn(HEDTA)(H_2O)]^-$ is more stable than $MnNH_4PO_4 \cdot H_2O$ and does not react with the $(NH_4)_2HPO_4$ or, to put it another way, $MnNH_4PO_4 \cdot H_2O$ is soluble in aqueous EDTA solution because of the formation of a stable coordination compound.

SPECIFIC REFERENCES

1. R. C. CRIPPEN, "Determination of Manganese as Manganese Ammonium Phosphate Monohydrate", *Chemist-Analyst*, 29, 54 (1940).
2. M. F. C. LADD, "The Radii of Spherical Ions", *Theor. Chim. Acta*, 12, 333 (1968).
3. S. RICHARDS, B. PEDERSEN, J. V. SILVERTON, and J. L. HOARD, "Stereochemistry of Ethylenediaminetetraacetato Complexes. I. The Structure of Crystalline $Mn_3(HY)_2 \cdot 10\ H_2O$ and the Configuration of the Seven-Coordinate $Mn(OH_2)Y^{-2}$ Ion", *Inorg. Chem.*, 3, 27 (1964).

PROCEDURE

Before beginning the analysis, your sample to be analyzed should be dried in an oven at 110°C for 1 hour. Store it in a desiccator until you are ready to use it.

The analysis should be carried out in duplicate.

TABLE 22-2

Specific Directions as Indicated in Text

	Mg analysis	*Mn analysis*
Recommended fritted crucible porosity	Medium	Course
Sample size	0.75 - 0.85 g	0.60 - 0.80 g
Dissolving of sample	Dissolve in 75 ml of distilled water	Use 75 ml of distilled water, add 15 drops of 3 M HCl, and heat until dissolved
Amount of NH_4Cl	1.0 g	10.0 g
Formula of precipitate obtained from hot solution	$MgNH_4PO_4 \cdot H_2O$	$MnNH_4PO_4 \cdot H_2O$
Molecular weight of precipitate	155.32	185.97

Obtain two fritted glass crucibles (see Table 22-2 for recommended porosity) or prepare two Gooch crucibles with asbestos as described on p. 39. Decide how you will tell the two crucibles apart, place them in two separate small beakers labeled with your identification mark (initials or drawer number), and dry in a 110°C oven for 30 minutes. Allow them to cool in a desiccator for at least 20 minutes and weigh to the nearest 0.1 mg. Store them in a desiccator until you are ready to use them. (How did you handle the crucibles? When hot, tongs are necessary to keep from burning your fingers; when cool, tongs are still necessary to keep from contaminating the crucibles.)

Using weighing pans or glazed paper, weigh two samples to the nearest 0.1 mg, each within the range indicated in Table 22-2. Transfer these to two marked 250 ml beakers and dissolve (see Table 22-2). Heat to near 70°C and add to each, with stirring, solid NH_4Cl (see Table 22-2 for amount), 1.5 g of $(NH_4)_2HPO_4$, and 2 ml of 15 M aqueous ammonia in that order. Digest the precipitates by heating gently near the boiling point for 15 minutes. Allow them to cool; the cooling may be hastened by placing the beakers in ice. Transfer each precipitate quantitatively (see p. 24) to one of the previously weighed filter crucibles using a minimum amount of water from your wash bottle. (Are you keeping track of which sample goes in which crucible?) Add 15 ml of 1.5 M aqueous ammonia to the beaker, scrub with a rubber policeman to loosen any particles adhering to the glass, and transfer this (wash liquid and solid) to the crucible. Repeat with a second 15 ml portion of 1.5 M NH_3. Pour in succession three 10 ml amounts of 95% ethanol into the crucible and allow air to be sucked through the solid for 1 or 2 minutes after the last of the ethanol has passed through. Dry the crucibles and contents in a 110°C oven for 30 minutes and cool in a desiccator. When cool, weigh them to the nearest 0.1 mg.

EXERCISES

1. The analytical balance is capable of weighing to within ± 0.1 mg. If it is desired to carry out a determination to four significant figures, what is the minimum size sample that can be used?
2. Why is it necessary to heat the sample in an oven before starting the experiment?
3. (a) How many moles are there in 0.10 g of $MgSO_4$?
   (b) How many grams of $(NH_4)_2HPO_4$ are required to react with this sample ($MgNH_4PO_4 \cdot H_2O$ is the product)?
   (c) Compare this with the amount of $(NH_4)_2HPO_4$ used in this experiment.
4. In an analysis for magnesium, a student precipitated the magnesium as $Mg(C_9H_6NO)_2 \cdot 2H_2O$. The data are given below. Calculate the percent magnesium for each sample and the average percent magnesium. The molecular weight of $Mg(C_9H_6NO)_2 \cdot 2H_2O$ is 348.65.

Sample No.	Sample weight (g)	Precipitate weight (g)
1	0.5048	0.4965
2	0.2551	0.2512
3	0.4485	0.4427

5. In analyzing for chloride ion, a student in a triplicate analysis obtained 45.60%, 45.88% and 45.55%. Calculate the average percent chloride, the deviation of each measurement from the average, and the relative average deviation in parts per thousand.

(Answers are on p. 253)

REPORT A GRAVIMETRIC ANALYSIS

1. Sample being analyzed _____

2. | | No. 1 | No. 2 | No. 3 |
    |---|---|---|---|
    | Weight container + sample (g) | _____ | _____ | _____ |
    | Weight container (g) | _____ | _____ | _____ |
    | Weight sample (g) | _____ | _____ | _____ |

3. | | No. 1 | No. 2 | No. 3 |
    |---|---|---|---|
    | Weight crucible + $MNH_4PO_4 \cdot H_2O$ (g) | _____ | _____ | _____ |
    | Weight crucible (g) | _____ | _____ | _____ |
    | Weight $MNH_4PO_4 \cdot H_2O$ (g) | _____ | _____ | _____ |

4. Calculate the percent magnesium or manganese in each sample:

    Sample 1

    Sample 2

    Sample 3

5. Summary of results:

	Percent metal	Deviation from average
Sample 1	_____	_____
Sample 2	_____	_____
Sample 3	_____	_____
Average	_____	_____

Relative average deviation: _____ ppt

6. Comparison of manganese analysis with theoretical:

   (a) Calculate the total theoretical percent manganese in the complex $MN_3(HEDTA)_2 \cdot 10H_2O$.

   (b) Considering the discussion in the introduction, what is the theoretical percent manganese for analysis by the method used in this experiment?

   (c) Calculate your deviation from the theoretical percentage in (b) in parts per thousand.

7. Since the relative error in an analysis becomes smaller as the sample size is increased, why don't you weigh out samples twice as large as those specified in the experiment?

Name _____ Section _____ Grade _____

Experiment 23
Nitrogen Content of an Amino Acid

Compounds with "free" NH_2 groups (NH_2 groups bonded to only one carbon) react with nitrous acid to liberate nitrogen, N_2. If the NH_2 group is one carbon away from a carboxyl group, as in α-amino acids, the reaction is rapid at room temperature; other NH_2 groups react more slowly, sometimes requiring up to 8 hours for complete reaction.

$$R-NH_2\ (aq) + HNO_2\ (aq) \longrightarrow R-OH\ (aq) + H_2O\ (l) + N_2\ (g) \tag{1}$$

Experiment 13 has a brief discussion of amino acids and describes a laboratory method of preparation of glycine, an amino acid that is to be analyzed in this experiment.

In 1911, D. D. Van Slyke developed a method for the determination of the nitrogen content of biological materials containing free NH_2 groups based on Eq. (1). The sample to be analyzed is dissolved in an acetic acid solution and reacted with sodium nitrite. The $NaNO_2$ in an acid solution generates nitrous acid which reacts with the amino acid to liberate nitrogen gas which is collected and its volume measured. From this volume and the temperature and pressure of the gas, the amount of nitrogen collected can be calculated using the ideal gas equation.

Gases are frequently collected by the displacement of a liquid. Ideally, this liquid should have a negligible vapor pressure and should not react with the gas. The gas should not be appreciably soluble in the liquid. Mercury is the most common liquid used, but water is much easier to work with and will be used in this experiment. Water has an appreciable vapor pressure (see Table 23-1) and the gas above the water will not only contain the N_2

TABLE 23-1

Vapor Pressure of Water

Temperature (°C)	V.P. (torr)	Temperature (°C)	V.P. (torr)
20.0	17.5	25.0	23.8
20.5	18.1	25.5	24.5
21.0	18.7	26.0	25.2
21.5	19.2	26.5	26.0
22.0	19.8	27.0	26.7
22.5	20.4	27.5	27.5
23.0	21.1	28.0	28.3
23.5	21.7	28.5	29.2
24.0	22.4	29.0	30.0
24.5	23.1	29.5	30.9

being collected but also H_2O. Using Dalton's Law of Partial Pressure, the pressure due to the N_2 can be calculated from the total pressure (assumed to be the same as atmospheric

$$P_{total} = P_{N_2} + P_{H_2O} \tag{2}$$

pressure) and the vapor pressure of water at room temperature. Nitrogen does not react with water but it is slightly soluble in water (about 2 ml of N_2 per 100 ml of H_2O at room temperature and pressure). The solubility error will be less than 2 ml of N_2 per 100 ml H_2O because the water will already contain some dissolved air that is 80% N_2.

Solutions of nitrous acid are unstable and dissociate according to Eq. (3). The NO_2 formed does not escape as a gas but reacts with the water according to Eq. (4). Thus the dissociation of nitrous acid liberates nitrogen oxide gas, NO, which is not very soluble in water (about 5 ml per 100 ml of H_2O at room temperature and pressure).

$$2\ HNO_2\ (aq) \longrightarrow NO\ (g) + NO_2\ (g) + H_2O\ (l) \tag{3}$$

$$3\ NO_2\ (aq) + H_2O\ (l) \longrightarrow 2\ HNO_3\ (aq) + NO\ (g) \tag{4}$$

Thus the reactions given in Eq. (1), (3), and (4) occur at the same time and the gas collected would be a mixture of N_2, H_2O, and NO if the NO is not removed. It is necessary to remove the NO because there is no way of determining the NO content of the collected gas. An effective way to remove NO is to pass the effluent gases over solid iron(II) sulfate heptahydrate, $FeSO_4 \cdot 7\ H_2O$. The NO is absorbed by the iron(II) sulfate probably by the formation of a coordination compound between the iron(II) and the NO of the type $[Fe(NO)(H_2O)_5]$.

PROCEDURE

Set up the apparatus as shown in Fig. 23-1 (a). Use a 250 ml beaker with about 175 ml of distilled water in it. The drying tube should contain enough $FeSO_4 \cdot 7\ H_2O$ to fill the

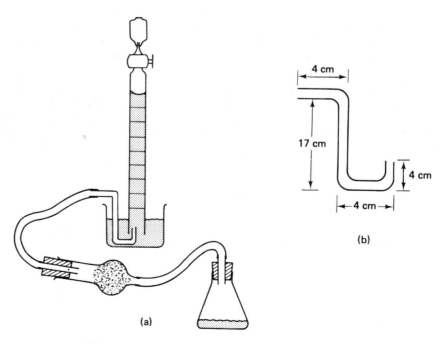

Figure 23-1. Apparatus for the collection of nitrogen

bulb and part of the neck of the drying tube; it is held in place by small pieces of glass wool. The dimensions of the gas delivery tube are shown in Fig. 23-1 (b). Prepare this tube from a 36 cm length of 6 mm soft glass tubing; instructions on glass bending are given on p. 43. Fill the buret to near the 50 ml mark with water by use of a pipet bulb attached to the buret as shown in Fig. 23-1 (a). If the level goes above the 50 ml mark, remove the bulb and lower the level by carefully opening the stopcock until the level is set. Record the buret reading to the nearest 0.01 ml remembering that the buret is upside down and so the numbers increase as you go toward the *top* of the buret.

To the nearest 0.1 mg weigh a 0.10 to 0.12 g sample of glycine using either the glycine prepared in Experiment 13 or a sample supplied by your instructor. If you weigh your sample by the indirect method (see p. 23), weigh it into a 50 ml Erlenmeyer flask and dissolve it in 7 ml of distilled H_2O. If you weigh your sample by the direct method using a weighing pan, use 7 ml of distilled water to transfer it to a 50 ml Erlenmeyer flask. The use of more water than this will give a more dilute solution causing the reaction to be significantly slower. Add 0.5 ml of concentrated acetic acid (CAUTION: This causes burns; wash with water immediately if any gets on your skin). The volume of acetic acid added is critical; it should be no more than 0.6 ml and no less than 0.3 ml.

When you are certain that the apparatus is ready, add 2 ml of 8 M $NaNO_2$ solution to the glycine solution and insert the stopper immediately. The gas evolution will begin in about 3 seconds and will continue for 1 to 2 minutes. After 2 minutes, shake the flask and if no more gas is bubbling into the buret, you may consider the reaction complete.

Record the buret reading to the nearest 0.01 ml, the room temperature, and the atmospheric pressure. Repeat the determination at least once and preferably twice.

CALCULATIONS

Calculate the percentage nitrogen in the glycine sample. It is suggested that you first calculate amount of nitrogen collected in the buret, using the ideal gas law. Keep Eq. (2) in mind when you do this calculation. Then calculate the weight of nitrogen present in your glycine sample. Keep Eq. (1) in mind when you do this calculation.

EXERCISES

1. A sample of helium gas was collected over water. The volume collected was 42.60 ml at 25.0°C and the atmospheric pressure was 750.8 torr. What weight of helium was collected?
2. A sample of sodium hydride was reacted with hydrochloric acid. If 0.825 mole of H_2 was collected, what weight of sodium hydride reacted?

$$NaH\ (s)\ +\ HCl\ (aq)\ \longrightarrow\ NaCl\ (aq)\ +\ H_2\ (g)$$

3. What is the minimum volume of 8 M $NaNO_2$ solution needed to react with 0.100 g of glycine? Compare this with the amount used in this experiment.
4. Calculate the theoretical percent nitrogen in phenylalanine, $C_6H_5CH_2CH(NH_2)COOH$.

(Answers are on p. 253)

REPORT NITROGEN CONTENT OF AN AMINO ACID

1. Room temperature (°C) _____
2. Atmospheric pressure (torr) _____
3. Vapor pressure of water at room temperature (torr) _____
4.   1   2   3

   Glycine + container (g) _____   _____   _____
   Container (g)   _____   _____   _____
   Glycine (g)   _____   _____   _____

5.   1   2   3

   Initial buret reading   _____   _____   _____
   Final buret reading   _____   _____   _____
   Volume of gas (ml)   _____   _____   _____

6. Calculate the weight of N_2 collected:

   1.

   2.

   3.

7. Calculate the percent nitrogen in the samples analyzed:
   1.

   2.

   3.

8. Summary of results:

	Percent N	Deviation from average
Sample 1	_____	_____
Sample 2	_____	_____
Sample 3	_____	_____
Averages	_____	_____

   Relative average deviation (ppt) _____

9. Calculate the theoretical percent nitrogen in glycine.

10. Assuming that your analysis is correct, calculate the percentage purity of the glycine.

Name _____ Section _____ Grade _____

Experiment 24
Spectrophotometric Analysis of Copper

There comes a time in the life of most scientists when it is necessary to do a quantitative analysis. The determination of the amount of a given chemical in a sample is a common occurrence in laboratories. If it is necessary to do a great many analyses, all of which are the same, an instrumental method is developed because such methods are rapid. Among the most widely used are optical methods, those involving absorption or emission of light. Even for nonroutine analyses, they are often rapid and reasonably accurate. This experiment, although not exactly the same as the procedures normally used for routine analyses, is quite similar and will serve as a good introduction to the general method of spectrophotometric analysis.

The applications of optical methods are varied and a few are mentioned here as examples. In biochemistry the coenzyme nicotinamide adenine dinucleotide (NAD), which takes part in metabolic oxidation-reduction reactions, may be determined in a reduced form by light absorption at 340 nm. Proteins may be analyzed by the Biuret determination where Cu^{2+} is added to form a colored complex with the protein (compare with this experiment). The standard method for analyzing detergents for phosphate content involves forming a yellow heteropoly ion containing phosphate, vanadate, and molybdate and measuring its light absorption at either 460 or 315 nm.

In this experiment, the copper content of a sample will be determined by complexing the Cu^{2+} with tetraethylenepentamine. The deep blue complex formed has its maximum absorption of visible light at 650 nm.[1]

$$NH_2 CH_2 CH_2 NHCH_2 CH_2 NHCH_2 CH_2 NHCH_2 CH_2 NH_2$$

tetraethylenepentamine

The Beer-Lambert law for the absorption of light by a substance is

$$A = abc \tag{1}$$

where A is the absorbance, a is the absorptivity (sometimes called the extinction coefficient), b is the length of the path the light travels through the sample, and c is the concentration of the absorbing material. The absorptivity is constant for a given compound at a given wavelength and b is constant if the same sample tube is used for all measurements. Some spectrophotometers measure the amount of light which passes through the sample (percent transmittance, %T) which may be converted to absorbance by the relation

$$A = - \log T \tag{2}$$

or

$$A = 2.000 - \log \%T \tag{3}$$

The usual procedure in doing a spectrophotometric analysis is to determine a calibration curve by measuring the absorbance at several different known concentrations. The absorbance of an unknown solution is then measured and its concentration determined from this calibration graph of concentration versus absorbance. Since we are only going to analyze one sample and since we know the concentration range where the Beer-Lambert law is valid (this was reported in Ref. 1), we can get by with a little less work.

Under conditions where the Beer-Lambert law is valid, absorbance is directly proportional to concentration. Applying this law to a solution of known copper ion concentration (designated by subscript s for standard), we can write

$$A_s = a_s b_s c_s \tag{4}$$

For a solution of unknown copper ion concentration (designated by subscript u for unknown), we can write

$$A_u = a_u b_u c_u \tag{5}$$

Dividing Eq. (5) by (4) we obtain

$$\frac{A_u}{A_s} = \frac{a_u b_u c_u}{a_s b_s c_s} \tag{6}$$

But since $a_u = a_s$ (the complex absorbing the light is the same in both solutions) and $b_u = b_s$ (the same sample tube is used for both solutions), Eq. (6) simplifies to

$$\frac{A_u}{A_s} = \frac{c_u}{c_s} \quad \text{or} \quad c_u = c_s \times \frac{A_u}{A_s} \tag{7}$$

SPECIFIC REFERENCE

1. T. B. CRUMPLER, "Tetraethylenepentamine as a Colorimetric Reagent for Copper", *Anal. Chem.*, **19**, 325 (1947).

PROCEDURE

Prepare a standard copper solution from the copper(II) chloride solution available in the laboratory (the concentration in grams per liter will be on the bottle). Pipet 10.00 ml of $CuCl_2$ stock solution into a clean (need not be dry) 100 ml volumetric flask (see p. 30-32.) Add 3 ml of 2% tetraethylenepentamine solution, mix by swirling, and dilute to the mark with distilled water. Mix the solution in the flask by repeated inversions (hold on to the stopper) and then transfer it to a stoppered flask for storage.

Into a small (50 or 100 ml) beaker, weigh to the nearest milligram a 0.20 to 0.22 g sample of your unknown. Add to the beaker 15 ml of distilled H_2O and 10 drops of 16 M HNO_3 and heat gently until the sample is dissolved. Neutralize the acid by adding

1 M NaOH dropwise with stirring until a precipitate forms that does not redissolve with stirring (What is this precipitate?). Add 4 ml of 2% tetraethylenepentamine solution, stir to mix, and allow it to stand for at least 2 minutes to permit the precipitate to dissolve. Transfer the solution to a clean (need not be dry) 100 ml volumetric flask (see p. 25), dilute to the mark with distilled H_2O and mix the solution.

Obtain two spectrophotometer tubes (approximately 1 cm across) from your instructor* with directions regarding the proper methods of cleaning and handling them. Mark one tube with an "S" and fill with the copper-tetraethylenepentamine solution; this is the sample tube. Mark the other tube with an "R" and fill with distilled H_2O; this is the reference tube. Prior to filling the tubes, you should rinse each tube several times with small portions of the solution that it will contain. The rinse solution is discarded. Finally, fill the tubes with the proper solution and wipe off the outside with lintless, nonabrasive tissue. Be sure that you do not place your fingers on the part of the tube through which the spectrophotometer beam will pass.

Instructions for the use of spectrophotometers are given on p. 47-50. Your instructor will give you specific instructions for the particular spectrophotometer in your laboratory. Set the wavelength at 650 nm (6500 Å) and do not change it during the course of the experiment. The instrument is adjusted to 0 %T and, with the water in the reference tube, to 100 %T. The reference tube is removed and the sample tube containing the standard copper solution is placed in the instrument. The percent transmittance (or absorbance) is read and recorded.

Replace the standard copper solution in the sample tube with the unknown copper solution (Did you rinse the tube thoroughly?) Reset the 0 %T and 100 %T as before, placing the sample tube into the instrument. Record the %T (or A) of the unknown solution.

The instrument is most sensitive between 20 and 60% T (0.70 and 0.22 absorbance). If your unknown solution did not fall into this range, weigh another sample of your unknown that will fall into this range. If you weigh out a larger sample, you may want to increase the amount of 16 M HNO_3 in proportion to the amount of sample. There is no need to change the amount of tetraethylenepentamine solution; the 4 ml is sufficient for any sample transmitting at least 20% and an excess does not cause an error.

When you are finished, rinse the spectrophotometer tubes with distilled water and return them to your instructor.

EXERCISES

1. Calculate the absorbance that corresponds to a percent transmittance of 49.7.
2. If a solution in a 1.00 cm cell has a concentration of 1.15×10^{-3} mole per liter and gives an absorbance of 0.411, what is the molar absorptivity?
3. 0.3933 g of $CuSO_4 \cdot 5H_2O$ is dissolved in enough water, using a volumetric flask, to give 1.000 liter of solution. What is the concentration in moles per liter of copper ion in this solution?
4. A solid sample weighing 0.1433 g is dissolved to give 100.0 ml of solution. If this solution contains 131.4 mg of Cu^{2+} per liter, what was the percent copper in the original sample?

(Answers are on p. 254)

---

*If you are using a Spectronic 20, two 13 × 100 mm test tubes may be used. They should be cleaned and rinsed with distilled H_2O.

REPORT SPECTROPHOTOMETRIC ANALYSIS OF COPPER

1. Unknown number _____

2. Stock $CuCl_2$ solution:

   Concentration in grams of $CuCl_2 \cdot 2H_2O$ per liter _____

   Concentration in moles of Cu^{2+} per liter _____

3. Standard copper solution:

   Concentration in moles of Cu^{2+} per liter _____

4. Weight of unknown sample:

	1	2
Weight container + unknown (g)	_____	_____
Weight container (g)	_____	_____
Weight unknown (g)	_____	_____

5.

	%T	Absorbance
Standard solution	_____	_____
Unknown solution No. 1	_____	_____
Unknown solution No. 2	_____	_____

6. Calculate the concentration (in moles per liter) of the unknown solution.

_____

7. Calculate the percent by weight of copper in the unknown.

_____

8. If the absorptivity of the copper-tetraethylenepentamine complex is 151 l/cm·mol, calculate the path length of your spectrophotometer tube.

9. This experiment assumes that the path length of your spectrophotometer tube is about 1 cm. How would you have to change the experiment if you were using tubes with a path length of about 2 cm?

Name _____ Section _____ Grade _____

Experiment 25
The Chromate-Dichromate Equilibrium

Chemists seem to believe that the topic of equilibrium is important. General chemistry texts devote 5 to 10% of their pages to the topic; courses in analytical chemistry emphasize the topic; and it puts in brief appearances in other chemistry courses. Chemists are not alone in this belief. The authors are occasionally cornered by members of the biology department who want their students to have a better understanding of equilibrium in aqueous solutions. The underlying reason is that an understanding of ionic equilibria (both qualitative and quantative) aids in understanding the chemical processes involved.

In this experiment you will measure the equilibrium constant for the formation of the dichromate ion, $Cr_2O_7^{2-}$, from hydrogen chromate ions, $HCrO_4^-$.

$$2\ HCrO_4^- \rightleftarrows Cr_2O_7^{2-} + H_2O \tag{1}$$

$$K = \frac{[Cr_2O_7^{2-}]}{[HCrO_4^-]^2} \tag{2}$$

These two ions will be the principal species if the hydrogen ion concentration is between 10^{-3} and 10^{-1} M. Equation (3) indicates that the hydrogen chromate ion is a weak acid[1] and in acid solution the position of equilibrium will lie to the left.

$$HCrO_4^- \rightleftarrows H^+ + CrO_4^{2-} \quad K_a = 3.20 \times 10^{-7} \tag{3}$$

$$H_2CrO_4 \rightleftarrows H^+ + HCrO_4^- \quad K_a = 1.21 \tag{4}$$

Chromic acid is a strong acid[2] and is essentially completely ionized to $HCrO_4^-$ as in Eq. (4). In the like manner, the hydrogen dichromate ion, $HCr_2O_7^-$, is not present in significant quantities because the position of equilibrium in Eq. (5) lies to the right.[2]

$$HCr_2O_7^- \rightleftarrows H^+ + Cr_2O_7^{2-} \quad K_a = 0.85 \tag{5}$$

Thus by carefully choosing our conditions we are able for all practical purposes to limit the types of ions in solution to the two involved in Eq. (1).

Both the $HCrO_4^-$ and the $Cr_2O_7^{2-}$ ions are colored due to absorption of visible light. The approximate absorption spectra in the region 190 to 400 nm (ultraviolet and part of the visible regions) are shown in Fig. 25-1. Notice that both ions absorb light at the same

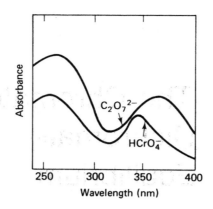

Figure 25-1. Absorption spectra of the dichromate and hydrogen chromate ions.

wavelengths but to differing degrees. If we make a measurement of the light absorbance (A_0) at a particular wavelength, we shall be measuring the sum of the absorbance of the $HCrO_4^-$ ion (A_1) and the $Cr_2O_7^{2-}$ ion (A_2).

$$A_0 = A_1 + A_2 \tag{6}$$

According to the Beer-Lambert law, absorbance $A = abc$ where a is the absorptivity (sometimes called the extinction coefficient), b is the length of the path the light travels through the sample, and c is the concentration of the absorbing material.

$$A_0 = a_0 b c_0, \quad A_1 = a_1 b c_1, \quad A_2 = a_2 b c_2$$

Substituting these expressions into Eq. (6) and solving for a_0 (notice that b cancels out since it is the same for all measurements), we obtain

$$a_0 = a_1 \frac{c_1}{c_0} + a_2 \frac{c_2}{c_0} \tag{7}$$

where c_0 is the total chromium(VI) concentration, c_1 is the $HCrO_4^-$ concentration, c_2 is the $Cr_2O_7^{2-}$ concentration, a_0 is the measured absorptivity, a_1 is the absorptivity of $HCrO_4^-$, and a_2 is the absorptivity of $Cr_2O_7^{2-}$.

In order to determine the equilibrium constant K in Eq. (2), we need to know c_1 and c_2 (c_0 and a_0 are obtained from our experimental data; a_1 and a_2 are obtained from Ref. 2). But Eq. (7) contains two unknowns so we need another relationship involving c_1 and c_2. Recall that all the chromium(VI) is present either as $HCrO_4^-$ or $Cr_2O_7^{2-}$. Putting that statement in the form of an equation gives

$$c_1 + 2c_2 = c_0 \tag{8}$$

If y is defined as the fraction of chromium(VI) atoms in the form of $Cr_2O_7^{2-}$ (and $1 - y$ is the fraction of chromium(VI) atoms in the form of $HCrO_4^{2-}$), then*

$$c_1 = (1 - y)c_0 \quad \text{or} \quad \frac{c_1}{c_0} = 1 - y \tag{9}$$

*In Eq. (10) the factor of one-half is included because c_0 is the total concentration of chromium(VI) and y times c_0 is the concentration of chromium(VI) in the form of dichromate ions. Since $Cr_2O_7^{2-}$ contains two chromium(VI) atoms per formula unit, the concentration of $Cr_2O_7^{2-}$ is one-half the chromium(VI) concentration.

$$c_2 = \tfrac{1}{2}yc_0 \quad \text{or} \quad \frac{c_2}{c_0} = \tfrac{1}{2}y \tag{10}$$

Substituting Eq. (9) and (10) into Eq. (7) gives

$$a_0 = a_1(1-y) + \tfrac{1}{2}a_2 y \tag{11}$$

Also substituting Eq. (9) and (10) into Eq. (2) gives

$$K = \frac{\tfrac{1}{2}yc_0}{[(1-y)c_0]^2} = \frac{y}{2(1-y)^2 c_0} \tag{12}$$

The approach in this experiment is to measure the absorbance of a chromium (VI) solution at 390 nm (A_0) and calculate the apparent absorptivity (a_0). Using this value and the reported absorptivities at 390 nm for $HCrO_4^-$ (a_1 = 519 l/cm·mol) and for $Cr_2O_7^{2-}$ (a_2 = 3480 l/cm·mol), the fraction of Cr(VI) as $Cr_2O_7^{2-}$ (y) is calculated. Then knowing y, the equilibrium constant K can be calculated. The references cited give values of the equilibrium constant as 43 (a glass electrode study[1]), 35.5, and 33.0 (spectrophotometric studies[2,3]). Compare your experimental value to these.

SPECIFIC REFERENCES

1. J. D. Neuss and W. Rieman, *J. Amer. Chem. Soc.*, **56**, 2238 (1934).
2. J. Y.-P. Tong and E. L. King, *J. Amer. Chem. Soc.*, **75**, 6180 (1953).
3. W. G. Davies and J. E. Prue, *Trans. Faraday Soc.*, **51**, 1045 (1955).

PROCEDURE

Instructions for the use of spectrophotometers are on pp. 47-50. Your instructor will give you specific directions for the particular spectrophotometer in your laboratory. Obtain two spectrophotometer tubes (approximately 1 cm across) from him* with instructions regarding the proper method of cleaning and handling them. Mark one tube with an "S"; this is the sample tube and will be filled with the solution to be measured. Mark the other tube "R"; this is the reference tube and should be filled with distilled water. Prior to filling the tubes, you should rinse each tube several times with small portions of the solution that it will contain. The rinse solution is discarded. Finally, fill the tubes with the proper solution and wipe off the outside with lintless, nonabrasive tissue. Be sure that you do not place your fingers on the part of the tube through which the spectrophotometer beam will pass. Set the wavelength at 390 nm (3900 Å) and do not change it during the course of the experiment. The instrument is adjusted to 0 %T and, with distilled water in the reference tube, to 100 %T. The reference tube is removed and the sample tube containing the solution to be measured is placed in the instrument. The percent transmittance (or absorbance) is read and recorded.

A. Determination of the path length of the spectrophotometer tube.

To the nearest 1 mg, weigh a 2.2 to 2.6 g sample of $NiCl_2 \cdot 6H_2O$. Transfer the sample to a clean (need not be dry) 100 ml volumetric flask (see pp. 23 and 25), dilute to the mark with distilled water, and mix thoroughly. Transfer the solution to a clean *dry*

---

*If you are using a Spectronic 20, two 13 × 100 mm test tubes may be used. They should be cleaned and rinsed with distilled H_2O.

flask for storage. Measure the percent transmittance (or absorbance) of this solution at 390 nm.

Applying the Beer-Lambert law to this solution and using an absorptivity at 390 nm of 4.94 l/cm·mol, calculate an effective path length for your spectrophotometer tube. If you are using a cylindrical tube, the effective path length may not be the same as the true inside diameter due to the round glass surfaces acting as a lens at this wavelength light.

B. Determination of apparent absorptivity of a chromium(VI) solution.

A solution about 10^{-3} M in chromium(VI) is needed for this experiment. In order to prepare such a dilute solution with sufficient accuracy, it is necessary to first prepare a more concentrated solution and then dilute it quantitatively. These solutions should also be about 10^{-2} M in H^+.

To the nearest 0.1 mg, weigh a 0.14 to 0.15 g sample of $K_2Cr_2O_7$. Transfer it to a clean (need not be dry) 100 ml volumetric flask, add 30 to 50 ml of distilled water, and swirl the flask until the solid is dissolved. Add 10 ml of 0.1 M HCl, dilute to the mark with distilled water, and mix thoroughly. Transfer the solution to a clean *dry* beaker. Rinse out the flask with distilled water (discard the washings) before going on with the dilution.

Pipet 10.00 ml of the $K_2Cr_2O_7$ solution into the 100 ml volumetric flask. Add 9 ml of 0.1 M HCl and dilute to the mark with distilled water. Mix the solution thoroughly and measure its percent transmittance (or absorbance) at 390 nm.

EXERCISES

1. Calculate the absorbance that corresponds to a percent transmittance of 55.8.
2. If a solution in a 1.00 cm cell has a concentration of 2.68×10^{-3} mol per liter and gives an absorbance of 0.484, what is the absorptivity?
3. 0.8848 g of ammonium heptamolybdate tetrahydrate, $(NH_4)_6Mo_7O_{24} \cdot 4H_2O$ (M.W. = 1235.9), is dissolved in enough water using a volumetric flask to give 250.0 ml of solution.
   (a) What is the concentration of ammonium heptamolybdate in the solution?
   (b) What is the concentration expressed as moles of Mo(VI) per liter of solution?
4. Check on the statement made in the introduction to this experiment that the concentration of chromate ion in the acidity region used is negligible. Assume a pH of 2.00, a total chromium(VI) concentration of 1.0×10^{-3} M, and ignore any formation of dichromate ions.

(Answers are on p. 254)

REPORT THE CHROMATE-DICHROMATE EQUILIBRIUM

1. Container + $NiCl_2 \cdot 6H_2O$ (g) _____

    Container (g) _____

    $NiCl_2 \cdot 6H_2O$ (g) _____

2. Calculate the concentration of the $NiCl_2$ solution in moles per liter.

3. Percent transmittance _____ Absorbance _____

4. Using an absorptivity of 4.94 l/cm·mol, calculate the effective path length of your spectrophotometer tube in centimeters.

5. Container + $K_2Cr_2O_7$ (g) _____

    Container (g) _____

    $K_2Cr_2O_7$ (g) _____

6. Calculate the concentration of the $K_2Cr_2O_7$ solution in moles per liter.

    (a) Before dilution (the first solution)

    (b) After dilution (the solution used for measurement)

7. Calculate the concentration of the diluted $K_2Cr_2O_7$ solution in terms of moles of Cr(VI) per liter.

8. Percent transmittance _____ Absorbance _____

9. Applying the Beer-Lambert law, calculate the apparent absorptivity of the Cr(VI) solution.

10. Using Eq. (11), calculate the value of y, the fraction of chromium(VI) atoms present as dichromate ion.

11. Calculate the equilibrium constant for the dimerization of hydrogen chromate ions.

Name _____ Section _____ Grade _____

Experiment 26
Dissociation of an Iron(II) Complex

The rate at which phenanthroline is replaced by water molecules in a solution of the complex ion *tris*(1,10-phenanthroline)iron(II), $[Fe(phen)_3]^{2+}$, is easily measured. This rate of replacement will be determined by following the change in the optical properties of the bright orange complex solution.

1,10-phenanthroline, $C_{12}H_8N_2$, is a large, aromatic organic molecule that behaves as a bidentate ligand, forming stable complexes with numerous metal ions. The Fe(II) complex has the structure shown in Fig. 26-1, where the three 1,10-phenanthroline molecules are arranged around the Fe(II) ion in an octahedral geometry.

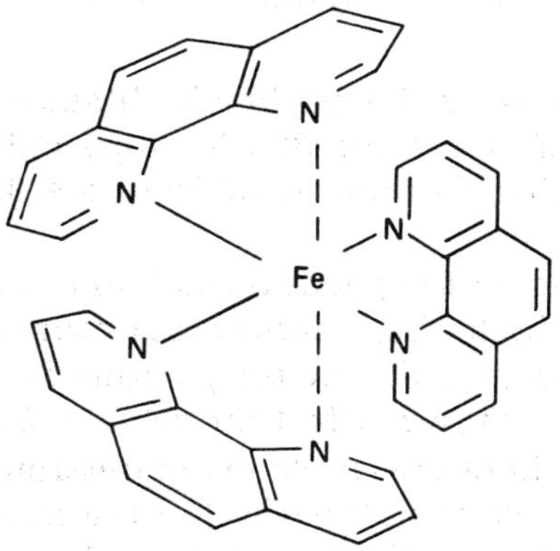

Figure 26-1. Structure of the $[Fe(phen)_3]^{2+}$ ion.

This complex ion is stable in neutral or weakly acidic solutions but dissociates in either basic or strongly acidic solutions. The reactions involved in the dissociation of the complex in acid solution are as follows:

(1) $\quad Fe(phen)_3^{2+} = Fe(phen)_2^{2+} + phen \quad$ slow reaction

(2) $\quad Fe(phen)_2^{2+} = Fe(phen)^{2+} + phen$

(3) $\quad Fe(phen)^{2+} = Fe^{2+} + phen \quad\quad\quad\quad$ fast reactions

As the phenanthroline molecules dissociate, water molecules fill the coordination positions around the iron.

Reactions (2) and (3) occur very rapidly but Reaction (1) proceeds slowly enough to be measurable and controls the rate of the overall reaction. It is this rate that will be measured in this experiment.

Reaction rates enable one to calculate the length of time required for a given reaction to occur under a specified set of conditions. In some cases, they provide an understanding of the pathways of the reaction and the existence and nature of intermediate species. It has been found experimentally that the reaction rate is dependent on the concentrations of one or more of the reacting substances. The simplest case is called *first-order kinetics*. It is expressed mathematically* as

$$\frac{dc}{dt} = -kc \qquad (1)$$

where c is the concentration of the reacting species, k is the specific reaction-rate constant, and dc/dt is the change in concentration with time (or the rate of the reaction). Rearranging Eq. (1) and integrating between limits gives

$$\int_{c_0}^{c} \frac{dc}{c} = -k \int_{0}^{t} dt \qquad (2)$$

$$\ln c - \ln c_0 = -kt \qquad (3)$$

where c_0 is the concentration of the reacting species at the beginning of the reaction ($t = 0$) and c is the concentration at some time (t) during the reaction. Rearranging Eq. (3) and converting to log base 10 gives Eq. (4).

$$\log c = -\frac{kt}{2.303} + \log c_0 \qquad (4)$$

When $\log c$ is plotted versus time (t), a straight line is obtained whose slope is $-k/2.303$ and whose intercept is $\log c_0$. Therefore $k = -2.303 \times$ slope and is determined experimentally in this manner. For a more detailed description of first-order kinetics you should consult your textbook.

Reaction rates are dependent on temperature, and kinetics experiments are usually done at constant temperature. You will do the experiment at room temperature and assume it to be constant during the laboratory period. The temperature is not needed to calculate the specific reaction-rate constants but it will affect the value of k obtained.

Numerous experimental methods are employed to determine the specific reaction-rate constant (k). The one chosen depends on the speed of the reaction under consideration and the physical and chemical properties of the compounds. An optical method has been chosen for this experiment because the $[Fe(phen)_3]^{2+}$ ion strongly absorbs light in the visible region of the spectrum, while the dissociation products absorb very little light in the same region. Therefore, as the reaction proceeds, the color of the solution fades, finally becoming colorless. This reaction is rather slow so that during the course of the laboratory period, the color change is barely noticeable to the naked eye. It is easily recorded, however, with the use of a spectrophotometer (see Fig. 26-2). Before performing the experiment you should read pp. 47-50 concerning spectrophotometers and the general principles of light absorption.

*If you have not taken calculus, the derivation beginning here and ending with Eq. (4) may have little meaning for you. It is still possible for you to use Eq. (4) which is of the form $y = mx + b$ where $y = \log c$, $m = -k/2.303$ and $b = \log c_0$.

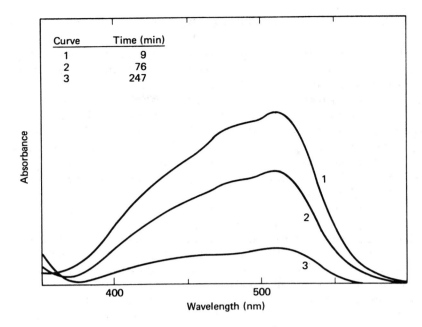

Figure 26-2. Spectrum of $[Fe(phen)_3]^{2+}$ solution at various times.

If one takes the Beer-Lambert law for the absorption of light by a substance

$$A = abc \tag{5}$$

where A is the absorbance, a is the absorptivity, b is the length of the path the light travels through the sample, and c is the concentration of the absorbing material, it can be seen that the absorbance is directly proportional to the concentration. The absorptivity is a constant for a given compound at a given wavelength and b is constant if the same sample tube is used for all measurements. Using Eq. (5) and substituting into Eq. (4) for c and c_0, one obtains

$$\log \frac{A}{ab} = -\frac{kt}{2.303} + \log \frac{A_0}{ab} \tag{6}$$

$$\log A = -\frac{kt}{2.303} + \log A_0 \tag{7}$$

where A_0 is the absorbance of the reacting species at the beginning of the reaction and A is the absorbance at some time during the reaction.

A plot of $\log A$ versus time will give a straight line whose slope equals $-k/2.303$. In this manner, the specific reaction-rate constant k can be obtained from the absorbance of the $[Fe(phen)_3]^{2+}$ solution.

Some spectrophotometers measure the amount of light that passes through the sample (percent transmittance, $\%T$), while others measure the amount of light absorbed by the sample (absorbance, A). They are related to each other by the expression

$$A = -\log T \tag{8}$$

or

$$A = 2.000 - \log \%T \tag{9}$$

These expressions may be used to convert one measurement into the other.

PROCEDURE

Pipet 10.00 ml of 0.001 M $FeSO_4$ stock solution into a clean (need not be dry) 100 ml volumetric flask (see p. 25.) Pipet 10.00 ml of 0.001 M 1,10-phenanthroline stock solution into the volumetric flask containing the $FeSO_4$ solution and mix them by swirling. At this concentration, formation of the ion, $[Fe(phen)_3]^{2+}$, can be assumed to be complete. This solution should be diluted with distilled H_2O until the flask is approximately three-fourths full and the solution mixed by gentle swirling. Add 5 ml of 1.0 M H_2SO_4 from a small graduated cylinder and record the time (t_0). Quickly dilute to the mark and mix the solution by inverting the stoppered flask numerous times.

Prepare a 0.05 M H_2SO_4 solution to be used in the reference tube by diluting 5 ml of 1.0 M H_2SO_4 to 100 ml with distilled water.

Obtain two spectrophotometer tubes (approximately 1 cm across) from your instructor* and directions for the proper methods of cleaning and handling them. Mark one of the tubes with an "S". This is the sample tube and will be filled with the $[Fe(phen)_3]^{2+}$ solution. Mark the other tube with an "R". This will be the reference tube and will be filled with the 0.05 M H_2SO_4. Prior to filling the tubes, you should rinse each tube several times with small portions of the solution that it will contain. The rinse solution is discarded. Finally, fill the tubes with the proper solution and wipe off the outside with lintless, nonabrasive tissue. Be sure that you do not place your fingers on the part of the tube through which the spectrophotometer beam will pass.

Take your tubes to the instrument you will be using. Your instructor will give you specific directions for the particular spectrophotometer in your laboratory. Set the wavelength at 510 nm (5100 Å) and try to avoid changing it during the course of the experiment. The instrument is adjusted to 0% T and to 100% T with the 0.05 M H_2SO_4 solution in the spectrophotometer. The reference tube is removed and the sample tube containing the $[Fe(phen)_3]^{2+}$ solution is placed in the instrument. The percent light transmitted (or absorbance) is read and recorded along with the time. Take a reading every 10 to 15 minutes, recording the %T to the nearest 0.1% and the time for each. The 0 %T and 100 %T settings should be checked using the reference solution before each reading is taken. The sample tube should be removed from the spectrophotometer immediately after a reading and kept in a place away from drafts or heat. Record the room temperature. To control the temperature more carefully, the reference and sample tubes may be kept in a beaker of H_2O that is at room temperature. In this case, record the temperature of the H_2O. The tubes must be wiped dry before inserting them in the spectrophotometer.

Continue taking measurements as long as time permits; try to obtain at least six readings. When you are finished with the experiment, clean out the spectrophotometer tubes by rinsing them with distilled H_2O, and return them to your instructor.

CALCULATIONS

Convert %T to A (absorbance) according to Eq. (9). Plot log A versus time and draw the best straight line through your points. Determine the slope of the line and calculate k, the rate constant.

The value of k given in the literature is 4.5 × 10^{-3} min^{-1} at 25°C.[1] If you wish to determine the literature value of k at your experimental temperature, you may use the following data:[2]

°C	k (min^{-1})
17.5	1.02 × 10^{-3}
25.3	4.43 × 10^{-3}
34.6	22.2 × 10^{-3}

*If you are using a Spectronic 20, two 13 × 100 mm test tubes may be used. They should be clean and rinsed with distilled H_2O.

A plot of log k versus $1/T$ (°K) gives a straight line from which the value of k may be taken for any given temperature.

SPECIFIC REFERENCES

1. T. S. Lee, I. M. Kolthoff, and D. L. Leussing, *J. Amer. Chem. Soc.*, **70**, 3596 (1948).
2. F. Basolo, J. C. Hayes, and H. M. Neumann, *J. Amer. Chem. Soc.*, **76**, 3807 (1954).

EXERCISES

1. Calculate log 0.550.
2. Calculate the absorbance that corresponds to a percent transmittance of 35.0.
3. Plot the following data and determine the slope of the line.

Sucrose concentration (in g/100 ml)	Angle of rotation (in degrees)
5.18	3.35
10.30	6.64
15.29	10.26
20.30	13.31

4. Show how Equation (7) is obtained from Equation (6).

(Answers are on p. 255)

REPORT DISSOCIATION OF AN IRON(II) COMPLEX

1. Enter your data directly in the following table and perform the necessary calculations.

Time (t)	$t - t_0$ (min)	Percent transmittance	Absorbance (A)	log A
$t_0 =$				

Experimental temperature _____ °C

2. Plot the data as described in the calculation section and determine your value of k, the specific reaction-rate constant.

$k =$ _____

3. Plot the log k versus $1/T$ data given on p. 226. From your graph, determine the rate constant at 20.0 and 20.5°C.

k at 20.0°C _____ k at 20.5°C _____

What is the percentage change in the rate constant going up in temperature by 0.5 from 20.0°C?

Name _____ Section _____ Grade _____

Experiment 27
Hydrolysis of a Tertiary Halide

Tertiary alkyl halides will react with water, some solvents, and negatively charged species to effect a substitution of the halogen atom with another group. The general structure for a tertiary alkyl halide is

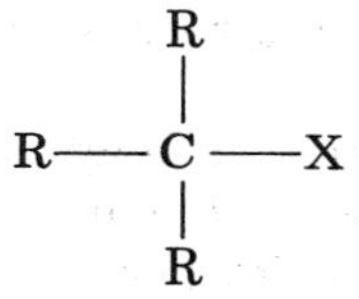

$$R-\underset{\underset{R}{|}}{\overset{\overset{R}{|}}{C}}-X$$

where X = F, Cl, Br, I
R = a carbon-containing group.

t-Butyl chloride and 1-bromoadamantane are two specific examples of tertiary alkyl halides (Fig. 27-1).

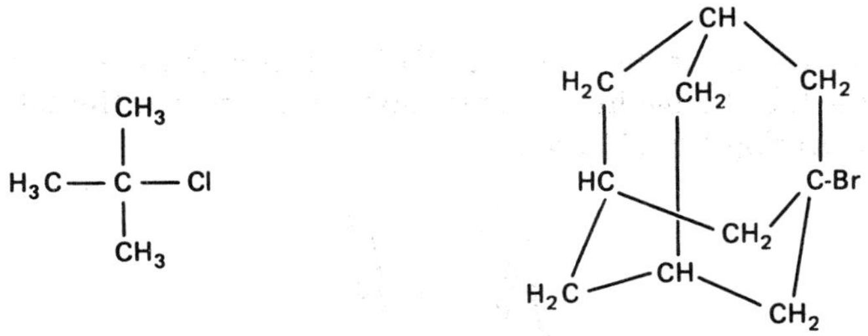

(a) $H_3C-\underset{\underset{CH_3}{|}}{\overset{\overset{CH_3}{|}}{C}}-Cl$

(b) 1-bromoadamantane structure

Figure 27-1. Tertiary halides.

Compounds of this type undergo substitution reactions; i.e., hydrolysis with water by a mechanism in which the first and rate-determining step is the ionization of the tertiary alkyl halide. The second step occurs very rapidly and does not affect the rate.

The rate of reactions of this type is dependent only on the concentration of the tertiary alkyl halide and the reaction is said to be *first order*. The concentration of the other reactant does not affect the rate. The reaction solvent does influence the rate, however,

(1) 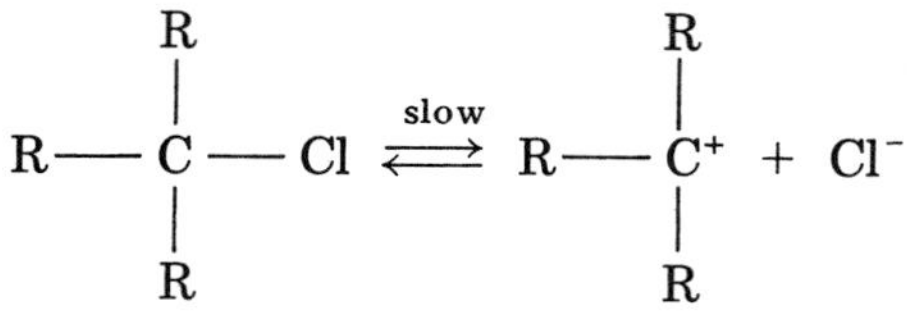 $$R-\underset{\underset{R}{|}}{\overset{\overset{R}{|}}{C}}-Cl \underset{}{\overset{slow}{\rightleftharpoons}} R-\underset{\underset{R}{|}}{\overset{\overset{R}{|}}{C^+}} + Cl^-$$

(2) 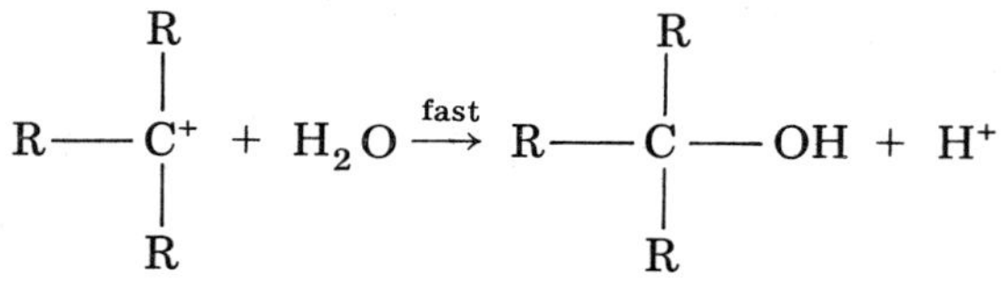 $$R-\underset{\underset{R}{|}}{\overset{\overset{R}{|}}{C^+}} + H_2O \overset{fast}{\rightarrow} R-\underset{\underset{R}{|}}{\overset{\overset{R}{|}}{C}}-OH + H^+$$

because the rate of ionization of the tertiary alkyl halide varies from one solvent to another. For example, the rate of hydrolysis of *t*-butyl chloride in alcohol-water solutions increases as the percentage of water in the solution increases. This occurs because water is a better ionizing medium for the tertiary alkyl halide than alcohol.

The hydrolysis of 1-bromoadamantane, Ad-Br, will be studied in this experiment. The solvent system of 40% ethanol was chosen because it affords a reaction rate that can be measured easily in the duration of a laboratory period. 1-Bromoadamantane has a rigid and highly sterically hindered structure and, therefore, cannot undergo side reactions that can occur to a small extent with many other tertiary alkyl halides. Its structure is a bit unusual and it certainly would not be classified as a "garden-variety" organic compound. Considerable research is being done in adamantane chemistry since many derivatives of adamantane have been shown to have antiviral activity particularly toward specific strains of influenza. The pharmacological efficacy appears to be due to the fact that the adamantyl group is a large hydrocarbon that reacts in a very specific manner and is resistant to metabolic attack.

Reaction rates enable one to calculate the length of time required for a given reaction to occur under a specified set of conditions. In some cases, they provide an understanding of the pathways of the reaction and the existence and nature of intermediate species. It has been found experimentally that a reaction rate is dependent on the concentration of one or more of the reacting substances The simplest case is called *first-order kinetics* and in this case the reaction rate is dependent on the concentration of just one of the substances. It is expressed mathematically* as

$$-\frac{dc}{dt} = kc \qquad (1)$$

where c is the concentration of the reacting substance, k is the specific reaction-rate constant, and dc/dt is the change in concentration with time (or the rate of the reaction). Rearranging and integrating Eq. (1) gives

$$-\int_{c_1}^{c_2} \frac{dc}{c} = k \int_{t_1}^{t_2} dt \qquad (2)$$

$$-\ln\left(\frac{c_2}{c_1}\right) = k(t_2 - t_1) \qquad (3)$$

or

$$-\log\left(\frac{c_2}{c_1}\right) = k\left(\frac{t_2 - t_1}{2.303}\right) \qquad (4)$$

*If you have not taken calculus, the derivation beginning here and ending with Eq. (4) may have little meaning for you. It is still possible for you to use Eq. (4) and the derivation of Eq. (5) from that.

If t_1 is chosen to be zero, then $c_1 = c_0$ or the concentration at the beginning of the reaction; if t_2 is chosen to be some time t, then $c_2 = c$, and Eq. (4) becomes

$$-\log\left(\frac{c}{c_0}\right) = \log\left(\frac{c_0}{c}\right) = \frac{kt}{2.303} \quad (5)$$

For a more detailed description of first-order kinetics, you should consult your textbook.

Reaction rates are dependent on temperature, and kinetic experiments are usually done at constant temperature. You will hydrolyze 1-bromoadamantane at room temperature, which is assumed to be constant throughout the laboratory period. Although the temperature does not enter into the calculation for the specific reaction-rate constant, it affects the value obtained.

The experimental method chosen to determine k depends on the speed of the reaction and the physical and chemical properties of the substances. In the hydrolysis of 1-bromoadamantane, 1 mole of hydrogen ions is generated for every mole of Ad-Br that reacts.

$$\text{Ad-Br} + H_2O \longrightarrow \text{Ad-OH} + H^+ + Br^-$$

The rates of Ad-Br consumption and H^+ formation are equal and the reaction can be followed simply by measuring the rate of H^+ production with a pH meter. In Eq. (5) the ratio c/c_0 represents the fraction of unreacted Ad-Br at time t. Because of the stoichiometry of the reaction this is also equal to the fraction of H^+ not yet formed at the same time t. Therefore it is possible to express the ratio of c/c_0 in terms of H^+ "concentrations."

$$\frac{c}{c_0} = (H_\infty^+ - H_t^+)/(H_\infty^+ - H_0^+) \quad (6)$$

where H_∞^+ is the hydrogen ion "concentration" at the end of the reaction, H_0^+ is the hydrogen ion "concentration" at the beginning of the reaction, and H_t^+ is the hydrogen ion "concentration" at time t. Substituting Eq. (6) into (5) gives

$$\log \frac{(H_\infty^+ - H_0^+)}{(H_\infty^+ - H_t^+)} = \frac{kt}{2.303} \quad (7)$$

When $\log (H_\infty^+ - H_0^+)/(H_\infty^+ - H_t^+)$ is plotted versus time, a straight line should be obtained whose slope is $k/2.303$. In this way you will determine the specific reaction-rate constant for the hydrolysis of 1-bromoadamantane. It must be noted here that the value that is measured by the pH meter is not really a hydrogen ion concentration or hydrogen ion activity because of the nonaqueous nature of the solvent system. The value is related, however, to the hydrogen ion concentration by a factor that is constant in the concentration range of this experiment. In doing the calculations the constant will cancel out and therefore we need not be concerned about it here.

It is not possible to measure directly the hydrogen ion "concentration" at the time of mixing H_0^+ because hydrolysis begins as soon as the aqueous solvent is added. Furthermore, it cannot be assumed that this value of H_0^+ is the same as the pure solvent because some of the 1-bromoadamantane may have hydrolyzed in the jar from atmospheric moisture or from moisture in the absolute ethanol used to prepare the 0.01 M solution. Therefore we shall take the time of the first measurement as t_0 and the corresponding pH as pH_0. It is possible to do this because we are interested in the *changes* which occur during a certain period of time which does not necessarily need to include the first few minutes after mixing.

Before performing the experiment you should read pp. 46-47 concerning the operation of the pH meter.

PROCEDURE

At the very beginning of the period, fill a large beaker, ice bucket, or other large container with tap water and set it aside so that the water will be at room temperature by the end of the period.

Note: Two different ethanol, C_2H_5OH, and water solutions will be used in this experiment. Their concentrations are expressed as percent ethanol by volume. Be certain to use the specified concentration.

The electrodes of the pH meter must equilibrate in a 40% ethanol solution for at least 15 minutes prior to use for kinetic measurements. Check the electrodes of the pH meter you will be using and place them in 40% ethanol if necessary. All pH measurements should be recorded to two decimal places. If the pH meter scale is marked in divisions of 0.1 pH unit, then it will be necessary to estimate between the divisions to the nearest 0.01 pH unit.

Obtain a clean 100 ml volumetric flask. If it is wet with water, rinse it with a few milliliters of acetone (CAUTION: flammable) and let it air-dry. It must be *completely dry* before it can be used. Pipet 10.00 ml of a freshly prepared 0.01 M solution of 1-bromoadamantane in absolute ethanol into the volumetric flask.

All your glassware and equipment should be ready and waiting before proceeding further with the reaction solution. Record the temperature of the room. It is assumed that all the chemicals and solutions that you are using are at room temperature so that this is the temperature at which the reaction will occur.

Dilute the 1-bromoadamantane solution in the 100 ml volumetric flask to the mark with 33% ethanol. Since the 1-bromoadamantane was dissolved in absolute ethanol, this dilution results in a 40% solution. Note the time and record it as t_{mix}. Mix the solution by repeatedly inverting the flask. It is desirable that the first pH measurement be taken as soon as possible after the solution has been mixed; e.g., within 5 minutes or less of t_{mix}. Pour all the diluted 1-bromoadamantane solution into a clean, *dry*, 150 ml beaker containing a magnetic stir bar. Place the pH electrodes in the solution, *being certain to allow sufficient clearance for the magnetic stir bar.* It would be wise to place an asbestos pad or other insulating material under the beaker to prevent heat transfer from the stirring motor to the reaction solution. Wrap a piece of Parafilm or plastic wrap (must be alcohol insoluble) over the top of the beaker and around the electrodes to prevent solvent evaporation. Do not remove the electrodes from the solution until the experiment is completed.

Now you should be ready to begin taking measurements. Stirring the solution *slowly*, record the time at t_0 and the pH at pH_0 for the first measurement. Continue recording the time and pH every minute for the first 10 minutes from t_0. As the rate of change in pH becomes smaller, you may increase the length of time between measurements, taking a total of 15 measurements during the experiment. At the end of approximately 1 hour the rate of change in pH will be quite small and you may stop taking measurements. After returning the pH meter to its "stand-by" setting, remove the electrodes and allow them to drain into the solution for 1 to 2 minutes. Then return the electrodes to the 40% ethanol soaking solvent.

It is still necessary to determine a value for the pH of the solution when the reaction has gone to completion, pH_∞. This value can be obtained by measuring the pH again after a long period of time has elapsed. It can be obtained in a much shorter period of time, however, by heating the solution and thereby greatly increasing the rate and essentially driving the hydrolysis reaction to completion. Tightly cover the beaker containing the 1-bromoadamantane solution with a piece of aluminum foil and place it in a beaker of hot water at 60°C for 10 minutes. Utmost care should be taken to prevent solvent loss, contamination with additional water, or anything that would change the concentration of the solution. Remove the beaker of reaction solution from the hot water and place it in the container of water you set aside at the beginning of the period. The reaction beaker should remain in the room temperature bath until it has reached room temperature also, roughly 15 minutes. Remove it from the water bath and measure the pH. Record this value as pH_∞.

At the end of the laboratory period the pH meter electrodes should be placed in a beaker of distilled water.

Two possible sources of error need special mention. (1) Since reaction rates are temperature dependent an accurate determination of a rate constant requires careful temperature control. You are assuming that room temperature remains constant during the period of taking measurements, which is a reasonable assumption for the accuracy with which the rest of the measurements are made. (2) Errors sometimes arise when a calculation involves a subtraction. If a small number is subtracted from a large number, a given error in either number may only give rise to a small relative error in the difference. However, when a number is subtracted from another number of about the same magnitude, a small error in either one of the numbers can cause a large relative error in the difference. Thus, you might predict that the accuracy with which the pH is measured might be important. It is possible to check this out and you are asked to do this in the laboratory report (Question 3) for an error of 0.01 pH unit. Since the pH is changing, this size error might be of greater or lesser importance at different times during the reaction and so you are asked to make two checks, one near the beginning and one at the end of your series measurements.

CALCULATIONS

Convert pH readings to hydrogen ion "concentrations". Calculate the ratio $(H_\infty^+ - H_0^+)/(H_\infty^+ - H_t^+)$. Plot the log $(H_\infty^+ - H_0^+)/(H_\infty^+ - H_t^+)$ versus time, determine the slope of the best straight line, and calculate k, the rate constant.

The literature value[1] for k at 25°C in 40% ethanol is 1.21×10^{-4} sec^{-1}.

SPECIFIC REFERENCES

1. J. Delhoste, G. Lamaty, and P. Pajanacci, "Solvolysis of 1-Bromoadamantane in Ethanol-Water Mixtures", *C. R. Acad. Sci., Paris, Ser. C*, **266**, 1508 (1968).
2. D. J. Raber, R. C. Bingham, J. M. Harris, J. L. Fry, and P. v. R. Schleyer, "The Role of Solvent in the Solvolysis of *t*-Alkyl Halides", *J. Amer. Chem. Soc.*, **92**, 5977 (1970). This reference contains a table of k values for 1-bromoadamantane in various solvent systems.

GENERAL REFERENCES

J. D. Roberts, and M. C. Caserio, "Basic Principles of Organic Chemistry", W. A. Benjamin, Inc., New York, N.Y., 1965, pp. 292-315.
Topics in Current Chemistry, 18, "Chemistry of Adamantanes", Springer-Verlag, Berlin, 1971, Chap. VI.

EXERCISES

1. What is the hydrogen ion concentration that corresponds to a pH of 2.47?
2. What is log 2.29?
3. See Exercise 3 of Experiment 26.
4. The slope of a log $(H_\infty^+ - H_0^+/H_\infty^+ - H_t^+)$ versus time (in minutes) plot was found to be 1.67×10^{-2}. What is k, the specific rate constant (a) in min^{-1}, (b) in sec^{-1}?
5. What is the half-life of the reaction in Exercise 3?

(Answers are on p. 256)

REPORT HYDROLYSIS OF A TERTIARY HALIDE

1. Enter your data directly in the following table and perform the necessary calculations.

   Time of mixing, t_{mix}: _____

Time (t)	$t - t_0$ (min)	pH	(H^+)	$\dfrac{(H_\infty^+ - H_0^+)}{(H_\infty^+ - H^+)}$	$\log \dfrac{(H_\infty^+ - H_0^+)}{(H_\infty^+ - H^+)}$
t_0 =	0.0	pH_0 =	H_0^+ =	1.00	0.000
t_∞ =	—	pH_∞^+ =	H_∞^+ =	—	—

Experimental temperature _____ °C

2. Plot the data as described in the calculation section.

3. Choose a point from the data you took during the first 15 minutes and assume that the pH measurement was 0.01 pH unit too large (i.e., subtract 0.01 from your measured pH value). Also choose a point from your data in the vicinity of 60 minutes and make the same assumption.

Using these new values of the pH, repeat the calculations as you did in Question 1. Record your results in the following table.

$t - t_0$ (min)	pH	(H^+)	$(H_\infty^+ - H_0^+)/(H_\infty^+ - H^+)$	$\log (H_\infty^+ - H_0^+)/(H_\infty - H^+)$

Plot these points on the graph that you prepared in Question 2. Comment on the error in $\log (H_\infty^+ - H_0^+)/(H_\infty - H^+)$ that arises from the assumed error in pH. In what region of the graph do you have the most confidence in your data?

4. Draw the best straight line through your points, taking into consideration the results of your calculations in Question 3. Determine the value of k, the specific reaction-rate constant.

$k = $ _____

Name _____ Section _____ Grade _____

APPENDIXES

Appendix I
Answers to Exercises

ANSWERS TO EXERCISES *Measuring Volume by Counting Drops*

1. Density of water at 23.0°C = 0.9975 g/ml

$$\text{Volume} = \frac{1.0058 \text{ g}}{0.9975 \text{ g/ml}} = 1.008 \text{ ml}$$

2.

Number	Deviation from average	$\left(\dfrac{\text{Deviation from}}{\text{average}}\right)^2$
55.54	0.83	0.69
53.96	0.75	0.56
54.42	0.29	0.08
56.23	1.52	2.31
53.87	0.84	0.71
53.67	1.04	1.08
55.06	0.35	0.12
57.75	3.04	9.24
51.86	2.85	8.12
(a) = 54.71	(b) = 1.28	

   (c) $\sigma = (22.91/8)^{1/2} = (2.864)^{1/2} = 1.692$

3. The 87% confidence limits are $\pm 1.5\sigma$. Thus the range would be $33.55 \pm 1.5\sigma$ or 33.55 ± 4.95 or 28.60 to 38.50.

4.

Number	Deviation
25.33	0.06
25.43	0.16
25.06	0.21
(27.01)	(1.74)
25.27	0.14 (Averages without suspected value).

Since the suspected value has a deviation from the average of more than four times the average deviation, it can be discarded.

Or, applying the Q test at the 90% level,
$$d = 27.01 - 25.43 = 1.58$$
$$r = 27.01 - 25.06 = 1.95$$
$$Q = \frac{d}{r} = \frac{1.58}{1.95} = 0.810$$

The observed Q value (0.810) is larger than the tabulated value (0.68 from Table 4 on p. 14). The questionable value may be discarded. Notice that the observed Q value is also greater than the tabulated value for the 95% confidence level.

ANSWERS TO EXERCISES   *Calibration of Volumetric Glassware*

1. At 27.8°C the density of water is 0.9964 g/ml.
$$\frac{9.98 \text{ g}}{0.9964 \text{ g/ml}} = 10.01 \text{ ml}$$

2. Correction = true value − measured value
   14.90 ml − 15.00 ml = −0.10 ml
3. True value = measured value + correction
   25.00 ml + (−0.05 ml) = 24.95 ml
4. See *Use of Volumetric Ware in Experiments* on p. 34 where it explains why you should try to use as much of your buret as possible.

ANSWERS TO EXERCISES   *Melting Points*

1. Error = observed value − correct value
   −0.2° − 0.0° = −0.2°C
   The minus sign indicates that the observed value is lower than the correct value.

   Correction = correct value − observed value
   0.0° − (−0.2) = +0.2°C
   The correction is the amount that must be (algebraically) added to the observed value in order to obtain the correct value.
2. The boiling point is lowered by 0.037°C for each torr decrease in pressure. Thus the change in the boiling point is
   (0.037°C/torr) (735.0 torr − 760.0 torr) = −0.93°C
   The boiling point at 735 torr is
   (100.0° − 0.9°) = 99.1°C
3. (A) The observed thermometer reading will be too high. Since the temperature will be increasing rapidly, one would need to be quite fast to observe both the melting of the solid and the thermometer reading simultaneously.
4. (D) It may be either (B) or (C) and is less likely to be (A) but more information is necessary to be certain which it is.

ANSWERS TO EXERCISES   *Formula of a Hydrate*

1. Of the ions listed, only SO_4^{2-} forms a barium salt that is insoluble in water and in nitric acid. All the cations listed react with H_2S or NaOH except Na^+. Thus the compound is Na_2SO_4.
2. A pink color would give an indication of a manganese(II) salt. A solution of Mn(II) is

very pale pink and would appear colorless if the concentration were low. A pink precipitate with H_2S confirms Mn^{2+}. Of the anions listed, only Cl^- and SO_4^{2-} give silver salts that are insoluble in water and in nitric acid. Both Cl^- and SO_4^{2-} give no reaction with $3\,M\,H_2SO_4$ but Cl^- does react with $18\,M\,H_2SO_4$ (liberating HCl) and SO_4^{2-} does not. Thus the compound is $MnCl_2$.

3. $$\%\text{ weight loss} = \frac{\text{weight loss}}{\text{sample weight}} \times 100 = \frac{0.1534}{2.0526} \times 100 = 7.473\%$$

4. Weight of hydrated sample = 20.5046 − 18.5421 = 1.9625 g
   Weight loss = weight of water = 20.5046 − 19.4983 = 1.0063 g
   Weight of anhydrous sample = 19.4983 − 18.5421 = 0.9562 g
   Formula weight of $MgSO_4$ = 120.4, of H_2O = 18.02
   Moles of water in hydrated sample is

   $$\frac{1.0063}{18.02} = 5.584 \times 10^{-2} \text{ mole } H_2O$$

   Moles of $MgSO_4$ in hydrated sample is

   $$\frac{0.9562}{120.4} = 7.942 \times 10^{-3} \text{ mole } MgSO_4$$

   Moles of water per mole of $MgSO_4 = \dfrac{(5.584 \times 10^{-2})}{(7.942 \times 10^{-3})} = 7.031$
   Formula of hydrate is $MgSO_4 \cdot 7H_2O$.

ANSWERS TO EXERCISES   *The Chemistry of Nitric Acid*

1. 0.50 g of Mg (at. wt., 24.31) is $\dfrac{0.50\text{ g}}{24.3\text{ g/mol}} = 2.1 \times 10^{-2}$ mol

   Concentrated HNO_3 is 16 *M*.

   (a) According to the equation

   $$Mg + 4\,HNO_3 \longrightarrow Mg(NO_3)_2 + 2\,NO_2 + 2\,H_2O$$

   1 mole of Mg requires 4 moles of HNO_3; thus 2.1×10^{-2} mole of Mg requires 8.4×10^{-2} mole of HNO_3.

   $$\frac{(8.4 \times 10^{-2}\text{ mol})(1000\text{ ml/liter})}{16\text{ mol/liter}} = 5.3 \text{ ml}$$

   (b) According to the equation

   $$3\,Mg + 8\,HNO_3 \longrightarrow 3\,Mg(NO_3)_2 + 2\,NO + 4\,H_2O$$

   1 mole of Mg requires 8/3 moles of HNO_3; thus 2.1×10^{-2} mole of Mg requires 5.6×10^{-2} mole of HNO_3.

   $$\frac{(5.6 \times 10^{-2}\text{ mol})(1000\text{ ml/liter})}{3\text{ mol/liter}} = 19 \text{ ml}$$

Notice that the dilute HNO_3 requires a smaller number of moles of nitric acid due to the greater reduction of the HNO_3 but a greater volume due to the lesser concentration.

2. 0.50 g of Pb (at. wt., 207.2) is $\dfrac{0.50 \text{ g}}{207.2 \text{ g/mol}}$ = 2.4 × 10^{-3} mol. The reaction is analogous to Answer 1 (a). Thus 1 mole of Pb requires 4 moles of HNO_3 and 2.4 × 10^{-3} Mole of Pb requires 9.6 × 10^{-3} mole of HNO_3.

$$\frac{(9.6 \times 10^{-3} \text{ mol})(1000 \text{ ml/liter})}{16 \text{ mol/liter}} = 0.60 \text{ ml}$$

Note that although the weight used is the same as in Answer 1(a), the amount of lead (mole-wise) is less and so the amount of nitric acid required is correspondingly less.

3.
$$1 \text{ mole Ti} \longrightarrow 1 \text{ mole TiO}_2$$
$$47.90 \text{ g Ti} \longrightarrow 79.90 \text{ g TiO}_2$$

$$\frac{79.90 \text{ g TiO}_2}{47.90 \text{ g Ti}} \times 0.5000 \text{ g Ti} = 0.8340 \text{ g TiO}_2$$

4. In 5 ml there are approximately 100 drops.

$$\frac{1 \text{ drop}}{100 \text{ drops}} \times 100 = 1\%$$

5. $\Delta G = \Delta H - T \Delta S$

(a) At 25°C,

$$\Delta G = 7.31 \text{ kcal} - (298°K)(0.0158 \text{ kcal}/°K) = +2.59 \text{ kcal}$$

A positive sign for ΔG indicates that the reaction as written is not spontaneous or that the reverse reaction is spontaneous. Thus Ag_2O is stable at 25°C. At 300°C,

$$\Delta G = 7.31 \text{ kcal} - (573°K)(0.0158 \text{ kcal}/°K) = -1.74 \text{ kcal}$$

The negative value of ΔG indicates that Ag_2O spontaneously decomposes to its elements at this temperature.
(b) At equilibrium, $\Delta G = 0$ and $\Delta H = T \Delta S$.

$$T = \frac{7.31 \text{ kcal}}{0.0158 \text{ kcal}/°K} = 463°K \text{ (190°C)}$$

Above 190°C, it is not possible to synthesize Ag_2O from its elements because of the instability of Ag_2O in this temperature range.

ANSWERS TO EXERCISES *Paper Chromatography*

1. Ink is composed of several colored ingredients. Applying ink might result in a separation into its components and interfere with the desired tests.

2.

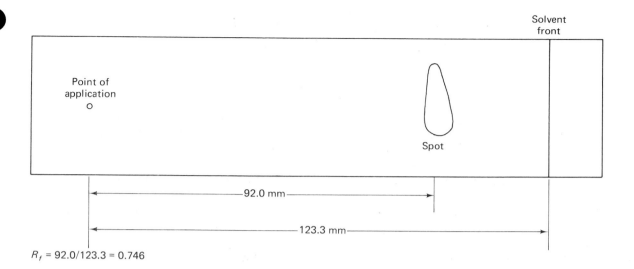

$R_f = 92.0/123.3 = 0.746$

3. The yellow precipitate must be CdS. If any one or more of the other ions were present, a black precipitate would have formed (mixed with the CdS). The negative tests indicate the absence of Pb^{2+}, Cu^{2+}, Co^{2+}, Ni^{2+}, and Fe^{3+},
4. The black precipitate could be any one or more of PbS, CuS, CoS, NiS, or FeS or it could be yellow CdS mixed with (and covered up by) one or more of those. The yellow precipitate with KI must be PbI_2, which confirms the presence of Pb^{2+} and the absence of Cu^{2+}. The negative test with SCN^- confirms the absence of both Co^{2+} and Fe^{3+}. The DMG test for Ni^{2+} may not be conclusive because the test must be carried out in slightly basic solution. If only one ion were present, it would be Pb^{2+} but if more than one were present, the presence or absence of Ni^{2+} has not been proved due to incorrect conditions for the test and the presence or absence of Cd^{2+} has not been proved because the yellow CdS may be covered up by the black PbS.

ANSWERS TO EXERCISES   *Thin-Layer Chromatography*

1.
$$R_f = \frac{77 \text{ mm}}{126 \text{ mm}} = 0.61$$

2.
$$\frac{77 \text{ mm}}{25.4 \text{ in/mm}} = 3.0 \text{ in}; \quad \frac{126 \text{ mm}}{25.4 \text{ in/mm}} = 4.96 \text{ in}$$

$$R_f = \frac{77/25.4}{126/25.4} = \frac{77}{126} = 0.61$$

As long as both distances are measured in the same units, the same R_f value will be obtained since the conversion factor cancels out when the ratio is taken.

3.
$$\frac{4 \text{ mg} \times \frac{1000 \text{ } \mu g}{\text{mg}}}{\text{ml} \times \frac{1000 \text{ } \mu l}{\text{ml}}} \times 3 \text{ } \mu l = 12 \text{ } \mu g$$

Note that 4 mg/ml is the same as 4 μg/μl.

$$\text{No. of moles} = \frac{12 \ \mu g}{10^6 \ \frac{\mu g}{g} \times 88.10 \ \frac{g}{mol}} = 1.4 \times 10^{-7} \ mol$$

ANSWERS TO EXERCISES *Ion-Exchange Chromatography*

1. Cu^{2+} and Co^{2+}: In aqueous solution, Cu(II) is blue and Co(II) is pink so together the solution will be slightly purple. Both CuS and CoS are black. Both $Co(OH)_2$ and $Cu(OH)_2$ are blue. The blue color of $[Cu(NH_3)_4]^{2+}$ is very intense and so initially it masks any color from the cobalt(II). As the Co(II) is oxidized to $[Co(NH_3)_6]^{3+}$, the color gradually changes $[Cu(SCN)_4]^{2-}$ is yellow-green and $[Co(SCN)_4]^{2-}$ is blue and together the solution appears yellow-green.
2. Cu^{2+}: In acetone-HCl solution, copper(II) is yellow. CuS is black. Fe(III) is also yellow and FeS is also black but the absence of a red color with NH_4SCN indicates the absence of iron(III).
3. Co^{2+}: This ion is colored but the intensity is low so that it doesn't always appear visible to the eye. CoS is black and $[Co(SCN)_4]^{2-}$ is blue.
4. For a given HCl concentration (0.5 M in this case) the greater the acetone concentration, the more stable the chloro anions of the metals are. Since this is a cation exchange column, the more stable the anions are made, the easier they are eluted from the column. Since the student was supposed to be adding an eluent with a high acetone concentration, he must have been eluting a metal that forms chloro anions with difficulty. Thus in using a lower concentration of acetone, he probably would not elute anything he didn't want to and so he can continue on with experiment; the only abnormality showing up is a portion of the eluate with no metal ion in it.
5. Go to jail! Go directly to jail! Do not pass GO! Do not collect $200! In other words, he had better start over again. The 90% acetone-0.5 M HCl will not only remove the metal ion he intended to remove with the 40% acetone-0.5 M HCl but other metal ions as well. Thus a separation will not be obtained and there is no way to reverse the mistake.

ANSWERS TO EXERCISES *Chemical Mystery Theater*

1. Solution 1 is $AgNO_3$; 2 is NaCl; 3 is NH_4OH; and 4 is HCl.
   The white precipitate is AgCl; AgCl is soluble in aqueous ammonia forming $[Ag(NH_3)_2^+]Cl^-$. Of the two solutions containing Cl^-, the NaCl solution would be neutral but the HCL solution would be acidic.
2. A is KBr; B is H_2S; C is $Mn(NO_3)_2$; D is K_2CrO_4; E is HCl; F is NaOH; G is Na_2CO_3.
   K_2CrO_4 solution is yellow and forms orange $K_2Cr_2O_7$ when acid is added. H_2S is the only candidate to have a vile odor. Na_2CO_3 liberates CO_2 (colorless, odorless gas) when reacted with acid. MnS and $Mn(OH)_2$ are both pink but the $Mn(OH)_2$ is easily oxidized by air to MnO(OH), which is brown. The KBr solution can only be identified by elimination of all other possibilities.

ANSWERS TO EXERCISES *Synthesis of Tin(IV) Iodide*

1. Atomic and molecular weights:

   Fe, 55.847; Cl_2, 70.906; $FeCl_3$, 162.21

   (a) $$\text{No. of moles Fe} = \frac{30.0 \text{ g Fe}}{55.85 \text{ g Fe per mole Fe}} = 0.537 \text{ mole Fe}$$

Since 1 mole of Fe yields 1 mole $FeCl_3$, 0.537 mole $FeCl_3$ is formed.

No. of grams $FeCl_3$ = 0.537 mole $FeCl_3$ × 162.2 g $FeCl_3$ per mole $FeCl_3$ = 87.1 g $FeCl_3$

(b) $\quad \%\text{ yield} = \dfrac{84.0}{87.1} \times 100 = 96.4\%$

2. Molecular weights:

$CrCl_3 \cdot 6H_2O$, 266.5; $\quad NaC_2H_3O_2$, 82.03; $\quad Cr(C_2H_3O_2)_2 \cdot H_2O$, 188.1

(a) From the reactions given, it can be seen that 1 mole of product comes from 1 mole of chromium(III) chloride and from 2 moles of sodium acetate.

No. of moles $CrCl_3 \cdot 6H_2O = \dfrac{29.9 \text{ g}}{266.5 \text{ g per mole}} = 0.112$ mole $CrCl_3 \cdot 6H_2O$

No. of moles $NaC_2H_3O_2 = \dfrac{84.0 \text{ g}}{82.03 \text{ g per mole}} = 1.02$ moles $NaC_2H_3O_2$

Since 1 mole of $CrCl_3$ reacts with 2 moles of $NaC_2H_3O_2$, then 0.112 mole of $CrCl_3$ would require 0.224 mole of $NaC_2H_3O_2$. The number of moles of $NaC_2H_3O_2$ available is much more than this so the limiting reagent is the $CrCl_3 \cdot 6H_2O$.

No. of grams $Cr(C_2H_3O_2)_2 \cdot H_2O$ formed =

0.112 mole $CrCl_3$ × $\dfrac{1 \text{ mole } Cr(C_2H_3O_2)_2 \cdot H_2O}{1 \text{ mole } CrCl_3}$ × 188.1 $\dfrac{g\ Cr(C_2H_3O_2)_2 \cdot H_2O}{\text{mole } Cr(C_2H_3O_2)_2 \cdot H_2O}$ =

21.1 g $Cr(C_2H_3O_2)_2 \cdot H_2O$

(b) $\quad \%\text{ yield} = \dfrac{18.0 \text{ g}}{21.1 \text{ g}} \times 100 = 85.3\%$

ANSWERS TO EXERCISES *Thermochemical Cycle*

1. 1.00 cal/g deg × 50.0 ml × 1.00 g/ml × 33.2 deg = 1660 cal
2. $K = \dfrac{(5424 - 5136) \text{ cal}}{12.84 \text{ deg}} = 22.4$ cal/deg
3. (a) Heat absorbed by the solution:

   1.00 cal/g deg × 50.0 ml × 1.00 g/ml × 22.3 deg = 1.12×10^3 cal or 1.12 kcal

   Heat absorbed by the calorimeter:

   2.68 cal/deg × 22.3 deg = 59.8 cal or 0.0598 kcal

   Total heat absorbed:

   1.12 + 0.06 = 1.18 kcal

   ΔH for the reaction = −1.18 kcal

(b) Moles Ca = $\dfrac{0.454 \text{ g}}{40.08 \text{ g/mole}} = 0.0113$ mole

(c) $\dfrac{-1.18 \text{ kcal}}{0.0113 \text{ mole}} = -104 \text{ kcal/mole}$

4. $\frac{1}{2} N_2 \ (g) + \frac{3}{2} H_2 \ (g) \xleftarrow{-Q_1} NH_3 \ (g)$

   $\searrow \qquad \nearrow$

   $NH_3 \ (s)$

$$0 = Q_2 + S + (-Q_1)$$

$$Q_2 = Q_1 - S = -11.04 - (4.13) = -15.17 \text{ kcal}$$

5. Convert Eq. (1) and twice the reverse of Eq. (3) to net ionic equations.

$$\text{Mg } (s) + 2 \text{ HCl } (aq) \longrightarrow \text{MgCl}_2 \ (aq) + \text{H}_2 \ (g)$$

$$\text{Mg } (s) + 2 \text{ H}^+ (aq) + 2 \text{ Cl}^- (aq) \longrightarrow \text{Mg}^{2+} (aq) + 2 \text{ Cl}^- (aq) + \text{H}_2 \ (g)$$

$$\text{Mg } (s) + 2 \text{ H}^+ (aq) \longrightarrow \text{Mg}^{2+} (aq) + \text{H}_2 \ (g) \qquad (5\text{-}1)$$

and

$$2 \text{ H}_2\text{O } (l) + 2 \text{ NaCl } (aq) \longrightarrow 2 \text{ HCl } (aq) + 2 \text{ NaOH } (aq)$$

$$2 \text{ H}_2\text{O } (l) + 2 \text{ Na}^+ (aq) + 2 \text{ Cl}^- (aq) \longrightarrow 2 \text{ H}^+ (aq) + 2 \text{ Cl}^- (aq) + 2 \text{ Na}^+ (aq) + 2 \text{ OH}^- (aq)$$

$$2 \text{ H}_2\text{O } (l) \longrightarrow 2 \text{ H}^+ (aq) + 2 \text{ OH}^- (aq) \qquad (5\text{-}2)$$

Summing these two net ionic equations [Eq. (5-1) and (5-2)] gives

$$\text{Mg } (s) + 2 \text{ H}_2\text{O } (l) \longrightarrow \text{Mg}^{2+} (aq) + 2 \text{ OH}^- (aq) + \text{H}_2 \ (g)$$

Converting this net ionic equation back to a molecular equation gives

$$\text{Mg } (s) + 2 \text{ H}_2\text{O } (l) \longrightarrow \text{Mg(OH)}_2 \ (aq) + \text{H}_2 \ (g)$$

since $\text{Mg(OH)}_2 \ (aq)$ is the same as $\text{Mg}^{2+} (aq) + 2 \text{ OH}^- (aq)$

Taking ΔH for Eq. (1) minus twice ΔH for Eq. (3) gives $\Delta H = -83.6$ kcal.

ANSWERS TO EXERCISES *Determination of a Simple Phase Diagram*

1. $\dfrac{5.00 \text{ g}}{80.00 \text{ g/mole}} = 0.0625 \text{ mole}$

2. Moles of $B = \dfrac{10.00 \text{ g}}{50.00 \text{ g/mole}} = 0.2000$

   Moles of $C = \dfrac{10.00 \text{ g}}{200.0 \text{ g/mole}} = 0.05000$

   Total moles $= 0.2000 + 0.05000 = 0.2500$

   Mole fraction of $B = \dfrac{\text{moles } B}{\text{total moles}} = \dfrac{0.2000}{0.2500} = 0.8000$

3. The compound is present in pure form at the maximum in the curve, which is at mole fraction of Mg of 0.33. Thus

$$X_{Mg} = 0.33 \quad \text{and} \quad X_{Zn} = 1.00 - 0.33 = 0.67$$

or

$$\frac{\text{moles Zn}}{\text{moles Mg}} = \frac{0.67}{0.33} = \frac{2}{1}$$

The compound is Zn_2Mg.

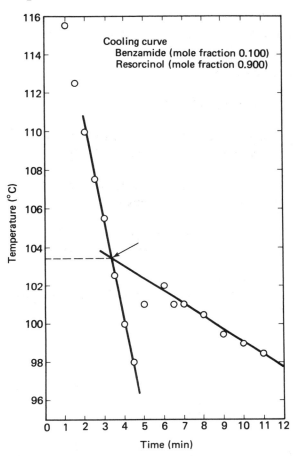

Figure 15-6.

4. 103.5°C. See graph of data in Fig. 15-6; melting point is at intersection of two lines.

ANSWERS TO EXERCISES *Heat of Fusion*

1. $$\frac{7.38 \text{ g}}{121.14 \text{ g/mole}} = 0.0609 \text{ mole}$$

2. $$\text{Moles of acetamide} = \frac{3.00 \text{ g}}{59.07 \text{ g/mole}} = 0.0508 \text{ mole}$$

$$\text{Moles of benzoic acid} = \frac{10.00 \text{ g}}{122.12 \text{ g/mole}} = 0.0819 \text{ mole}$$

$$\text{Total moles} = 0.0508 + 0.0819 = 0.1327 \text{ mole}$$

$$X_{\text{benzoic acid}} = \frac{\text{moles of benzoic acid}}{\text{total moles}} = \frac{0.0819}{0.1327} = 0.617$$

3. $\log 0.65 = \log (6.5 \times 10^{-1}) = 0.81 + (-1.00) = -0.19$
4. $\Delta H° = -\text{slope} \times 2.303 \times R$
   $= -(-763°\text{K})(2.303)(1.987 \text{ cal/°K·mol})(10^{-3} \text{ kcal/cal})$
   $= +3.49 \text{ kcal/mol}$
5. $\Delta S° = \text{intercept} \times 2.303 \times R$
   $= (2.16)(2.303)(1.987 \text{ cal/°K·mol}) = 9.88 \text{ cal/°K·mol}$

ANSWERS TO EXERCISES *Molecular Weight Determination by Freezing Point Depression*

1. $(1.034 \text{ g/ml})(17.45 \text{ ml}) = 18.04 \text{ g}$

2. $m = \dfrac{\text{No. of moles solute}}{\text{No. of kg solvent}} = \dfrac{(1.108 \text{ g})/(128.2 \text{ g/mole})}{(0.01500) \text{ kg}} = 0.5762 \, m$

3. $\Delta T_f = K_f \times \dfrac{(\text{g solute})/(\text{M.W. solute})}{\text{kg solvent}}$

   $\text{M.W.} = \dfrac{(\text{g solute})(K_f)}{(\Delta T_f)(\text{kg solvent})} = \dfrac{(2.91 \text{ g})(37.7 \text{ deg kg/mole})}{(178.4 - 151.4 \text{ deg})(0.0500 \text{ kg})} = 81.3 \text{ g/mole}$

4. As long as the thermometer error is consistent in the range being used, the error will cancel since the data being used is a *difference* between two temperatures.
5. See Exercise 4 of Experiment 15.

ANSWERS TO EXERCISES *An Acid-Base Titration*

1. $\text{Moles of KHP} = \dfrac{0.4415 \text{ g}}{204.2 \text{ g/mol}} = 0.002162 \text{ mol}$

   1 mole of KHP reacts with 1 mole NaOH

   $\text{Molarity of NaOH} = \dfrac{\text{moles of NaOH reacted}}{\text{liters of NaOH}} = \dfrac{0.002162 \text{ mol}}{0.02281 \text{ l}} = 0.09474 \, M$

2. Moles of NaOH used $= (0.1163 \text{ mol/liter})(0.02128 \text{ l}) = 0.002478 \text{ mol}$

   1 mole NaOH reacts with 1 mole acetic acid

   $\text{Molarity of HC}_2\text{H}_3\text{O}_2 = \dfrac{\text{moles of acid}}{\text{liters of acid}} = \dfrac{0.002478 \text{ mol}}{0.02500 \text{ l}} = 0.09900 \, M$

3. (a) Moles of NaOH $= (0.09939 \text{ mol/liter})(0.03575 \text{ l}) = 0.003553 \text{ mol}$
   1 mole NaOH reacts with 1 mole KHP
   g of KHP = mole KHP $\times$ 204.2 g/mol = $(0.003553 \text{ mol})(204.2 \text{ g/mol})$
   $= 0.7259$

   (b) $\% \text{ KHP} = \dfrac{\text{g of KHP}}{\text{sample weight}} \times 100 = \dfrac{0.7259 \text{ g}}{0.8070 \text{ g}} \times 100 = 89.95\%$

4. (a) Number of equivalents NaOH = (0.1222 equivalent/liter)(0.01894 liter)
$$= 0.002314 \text{ equivalent}$$
1 equivalent of NaOH reacts with 1 equivalent of acid
∴ number of equivalents of acid = 0.002314 equivalent

$$\text{Equivalent weight} = \frac{\text{No. of g acid}}{\text{No. of equivalents acid}} = \frac{0.1504 \text{ g}}{0.002314 \text{ equivalent}} = 65.00 \text{ g/equivalent}$$

(b) $\text{Equivalent weight} = \dfrac{\text{Molecular weight}}{\text{No. of protons neutralized}}$

Molecular weight = (Equivalent weight) (No. of protons)
$$= (65.00 \text{ g/equivalent}) (2 \text{ equivalents/mol}) = 130.0 \text{ g/mol}$$

5. When adding a solution dropwise from a buret, you can close the stopcock immediately after the drop falls and you then have added 1 drop. To add part of a drop, open the stopcock very gradually so that part of a drop is hanging on the buret tip. Then with water from your wash bottle, wash this part of a drop off the tip into the flask. By controlling the size of the drop hanging on the tip you can add either a very small fraction of a drop or almost a whole drop.

ANSWERS TO EXERCISES *Titration Using a pH Meter*

1. $$M_{\text{NaOH}} = \frac{(25.00 \text{ ml})(0.8360 \text{ }M)}{38.65 \text{ ml}} = 0.5408 \text{ }M$$

2. The molarity of the oxalic acid solution is 0.250 mol/0.250 l = 1.00 M.
   (a) Forming $HC_2O_4^-$ from $H_2C_2O_4$ is by neutralization of 1 proton so the normality is equal to the molarity: 1.00 N.
   (b) Forming $C_2O_4^{2-}$ from $H_2C_2O_4$ is by neutralization of 2 protons so the normality is twice the molarity: 2.00 N.
   (c) Forming CO_2 from $H_2C_2O_4$ is by loss of 2 electrons per oxalic acid molecule so the normality is twice the molarity: 2.00 N.

3. The end point obtained from the graph is 13.68 ml. For information on plotting data, see p. 6. The concentration of the base is 0.3149 N. Since no information is given about the base, it is not possible to calculate the concentration in terms of molarity.

ANSWERS TO EXERCISES *Volumetric Determination of Metals*

1. See p. 24.
2. (a) You have started with a definite weight of sample, which thus contains a definite weight of metal. The amount of water added does not change this weight of metal.
   (b) Keep in mind that the beaker holds about 250 ml and that you will be adding solution from your buret.
3. Some of the sample will be lost if solution remains on the electrodes.
4. The intense color of the Snazoxs will obscure the colors of other species for some metals, making it difficult if not impossible to detect the end point. In addition, the indicator also reacts with metal ion and must be displaced from the metal ion by the EDTA. If this displacement requires more than a very small volume of EDTA solution, the end point will not be sharp.
5.

24.35%	+0.25
24.05%	−0.05
23.90%	−0.20
3)72.30	3) 0.50
24.10%	0.17% average deviation

$$\text{rad} = \frac{0.17}{24.10} \times 1000 = 7.1 \text{ ppt}$$

6.
```
        18.90              −0.05
        19.00              +0.05
        18.95               0.00
     3 )56.85           3 ) 0.10
        18.95 ml            0.03 ml average deviation
```

$$\text{rad} = \frac{0.03}{18.95} \times 1000 = 1.6 \text{ ppt}$$

$$\text{Molarity of } Na_2H_2EDTA = \frac{(20.00)(0.009656)}{(18.95)} = 0.01019 \, M$$

7. 20 ml (0.020 l) of 0.010 M Na_2H_2EDTA contains 2.0×10^{-4} mole of EDTA. Since 1 mole of $EDTA^{4-}$ reacts with 1 mole of copper, then 2.0×10^{-4} mole of copper will be required.
Since 1 mole of enzyme contains 1 mole of copper, then 2.0×10^{-4} mole of enzyme will be required

$$(2.0 \times 10^{-4} \text{ mole})(5.60 \times 10^5 \text{ g/mole}) = 112 \text{ g}$$

This would be an unwieldy quantity to work with even if it were possible to obtain this much. The original analysis was done by atomic absorption and reported in *J. Biol. Chem.*, 244, 5761 (1969).

ANSWERS TO EXERCISES *Vitamin C Determination*

1. The molecular weight of ascorbic acid is 176.1:

$$(2.139 \times 10^{-5} \text{ mole}) (176.1 \text{ g/mole}) (1000 \text{ mg/g}) = 3.767 \text{ mg}$$

2. The total amount of ascorbic acid is the amount per aliquot times the number of aliquots in the total sample.

$$(16.01 \text{ mg/aliquot})(10 \text{ aliquots}) = 160.1 \text{ mg}$$

This amount was originally in 100 ml so the amount per milliliter is

$$\frac{160.1 \text{ mg}}{100 \text{ ml}} = 1.601 \text{ mg/ml}$$

3. $$\frac{6.17 \text{ mg}}{10.0 \text{ ml}} \times 29.57 \text{ ml/oz} \times 4.0 \text{ oz} = 73 \text{ mg}$$

4. $$\frac{22.72 \text{ ml}}{1000 \text{ ml/l}} \times 0.01274 \text{ mole/l} \times 176.1 \text{ g/mole} \times 1000 \text{ mg/g} = 50.97 \text{ mg}$$

5.
```
        7.208              0.024
        7.167              0.017
        7.178              0.006
     3 )21.553          3 )0.047
        7.184              0.016
```

$$\text{Relative average deviation} = \frac{0.016}{7.184} \times 1000 = 2.2 \text{ ppt}$$

ANSWERS TO EXERCISES *A Gravimetric Analysis*

1. 0.1000 g. A sample even 0.1 mg less than this (0.0999 g) would only be three significant figures.
2. Moisture content may vary slightly. Heating under controlled conditions removes this excess moisture; storing in a desiccator prevents the mosture from being readsorbed.
3. (a) $$\frac{0.10 \text{ g}}{120.4 \text{ g/mole}} = 8.3 \times 10^{-4} \text{ mole}$$

   (b) 1 mole $MgSO_4$ reacts with 1 mole $(NH_4)_2 HPO_4$; thus 8.3×10^{-4} mole of $(NH_4)_2 HPO_4$ is needed.

   $$(8.3 \times 10^{-4} \text{ mole})(132.1 \text{ g/mole}) = 0.11 \text{ g}$$

   (c) More than a ten fold excess of $(NH_4)_2 HPO_4$ is used in the experiment.

4. $$\% \text{ Mg} = \frac{\text{g of Mg in sample}}{\text{g of sample}} \times 100 = \frac{\text{moles of ppt.} \times \text{at. wt. Mg}}{\text{g of sample}} \times 100$$

   No. 1 $$\% \text{ Mg} = \frac{(0.4965/348.65)(24.31)}{0.5048} \times 100 = 6.858\%$$

   No. 2 $$\% \text{ Mg} = \frac{(0.2512/348.65)(24.31)}{0.2551} \times 100 = 6.866\%$$

   No. 3 $$\% \text{ Mg} = \frac{(0.4427/348.65)(24.31)}{0.4485} \times 100 = 6.882\%$$

   $$\text{Average} = \frac{6.858 + 6.866 + 6.882}{3} = 6.869\%$$

5.
   ```
        45.60           −0.08
        45.88           +0.20
        45.55           −0.13
     3 )137.03        3 ) 0.41
        45.68             0.14
   ```

   $$\text{Relative average deviation} = \frac{0.14}{45.68} \times 1000 = 3.1 \text{ ppt}$$

ANSWERS TO EXERCISES *Nitrogen Content of an Amino Acid*

1. Vapor pressure of water at 25.0°C = 23.8 torr
   Temperature = 273.2 + 25.0 = 298.2°K

   $$P_{He} = P_{total} - P_{H_2O} = 750.8 - 23.8 = 727.0 \text{ torr}$$

   $$n = \frac{PV}{RT} = \frac{(727.0 \text{ torr})(0.04260 \text{ liter})}{(62.36 \text{ l} \cdot \text{torr/mol} \cdot °K)(298.2°K)} = 1.665 \times 10^{-3} \text{ mole}$$

   No. of grams = $(n)(\text{at. wt.}) = (1.665 \times 10^{-3} \text{ mol})(4.003 \text{ g/mol}) = 6.667 \times 10^{-3}$ g

2. 1 mole of H_2 is formed from 1 mole of NaH; thus 0.825 mole of H_2 is formed from 0.825 mole of NaH.

$$0.825 \text{ mol NaH} \times \frac{24.00 \text{ g NaH}}{\text{mol NaH}} = 19.8 \text{ g NaH}$$

3. $$\text{Moles of glycine} = \frac{0.100 \text{ g}}{75.07 \text{ g/mole}} = 0.00133 \text{ mole glycine}$$

1 mole of glycine reacts with 1 mole of $NaNO_2$; thus 0.00133 mole of glycine reacts with 0.00133 mole of $NaNO_2$.

$$\frac{0.00133 \text{ mole } NaNO_2}{(8 \text{ mole of } NaNO_2/l)} = 0.000171 \text{ or } 0.17 \text{ ml}$$

More than a ten fold excess of $NaNO_2$ is used.

4. Molecular weight of phenylalanine is 165.19.

$$\% N = \frac{14.01}{165.19} \times 100 = 8.481\%$$

ANSWERS TO EXERCISES *Spectrophotometric Analysis of Copper*

1. $A = 2.000 - \log(\%T) = 2.000 - \log 49.7$
   $= 2.000 - 1.696 = 0.304$

2. $A = abc$

$$a = \frac{A}{bc} = \frac{0.411}{(1.00 \text{ cm})(1.15 \times 10^{-3} \text{ mol/liter})} = 357 \text{ l/cm·mol}$$

3. $$\text{Molarity} = \frac{\text{moles}}{l} = \frac{(0.3933 \text{ g})/(249.7 \text{ g/mol})}{1.000 \text{ l}} = 1.575 \times 10^{-3} \text{ M}$$

4. If 1 liter contains 131.4 mg of Cu^{2+}, then 100 ml will contain 13.14 mg or 0.01314 g of Cu^{2+}.

$$\% Cu = \frac{\text{g of copper}}{\text{g of sample}} \times 100 = \frac{0.01314}{0.1433} \times 100 = 9.170\%$$

ANSWERS TO EXERCISES *The Chromate-Dichromate Equilibrium*

1. $A = 2.000 - \log \%T = 2.000 - \log 55.8$
   $= 2.000 - 1.747 = 0.253$

2. $$a = \frac{A}{bc} = \frac{0.484}{(1.00 \text{ cm})(2.68 \times 10^{-3} \text{ M})} = 181$$

3. (a) $$M = \frac{(0.8848 \text{ g})/(1235.9 \text{ g/mole})}{(0.2500 \text{ l})} = 2.864 \times 10^{-3} \text{ M}$$

(b) 1 mole $(NH_4)_6Mo_7O_{24} \cdot 4H_2O$ contains 7 moles molybdenum(VI).

$$M = \frac{7 \text{ moles of Mo(VI)}}{\text{mole of } (NH_4)_6Mo_7O_{24}} \times \frac{2.864 \times 10^{-3} \text{ mole of } (NH_4)_6Mo_7O_{24}}{1} = 2.005 \times 10^{-2} \text{ M}$$

4.
$$HCrO_4^- \rightleftarrows H^+ + CrO_4^{2-}$$

$$K = \frac{[H^+][CrO_4^{2-}]}{[HCrO_4^-]} = 3.20 \times 10^{-7}$$

$[H^+] = 1.0 \times 10^{-2}$ since pH = 2.00

$[CrO_4^{2-}] = x$

$[HCrO_4^-] = 1.00 \times 10^{-3} - x$

$$\frac{(1.0 \times 10^{-2})(x)}{(1.00 \times 10^{-3} - x)} = 3.20 \times 10^{-7}$$

$$x = 3.2 \times 10^{-8} \, M$$

Percent of Cr(VI) as $CrO_4^{2-} = \dfrac{3.2 \times 10^{-8}}{1.00 \times 10^{-3}} \times 100 = 0.0032$

ANSWERS TO EXERCISES *Dissociation of an Iron(II) Complex*

1. 0.550 may be written as 5.50×10^{-1}
   $\log(5.50 \times 10^{-1}) = \log(5.50) + \log(10^{-1})$
   $\qquad\qquad\qquad\quad = (0.740) + (-1.000)$
   $\qquad\qquad\qquad\quad = -0.260$
2. $A = 2.000 - \log(\%T)$
   $\quad = 2.000 - \log 35.0 = 2.000 - 1.544$
   $\quad = 0.456$
3. The data is plotted on the following graph and the slope is determined from the points indicated by the dotted lines.

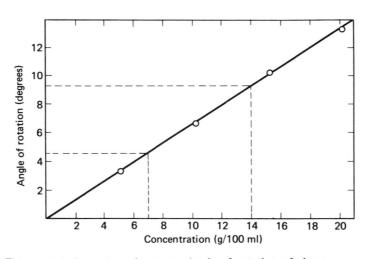

Figure 26-3. Inversion of sucrose. Angle of rotation of plane polarized light (5890 Å) as a function of concentration.

$$\text{Slope} = \frac{Y_2 - Y_1}{X_2 - X_1} = \frac{9.25 - 4.60}{14.00 - 7.00}$$

$$= \frac{4.65}{7.00} = 0.664$$

See p. 7 for additional information.

The slope in this case is the rate of change of the angle of rotation with respect to changes in sucrose concentration.

4. Equation (6) may be rewritten as follows since $\log(x/y)$ equals $\log x - \log y$.

$$\log A - \log(ab) = -\frac{kt}{2.303} + \log A_0 - \log(ab)$$

Since $-\log(ab)$ appears on both sides of the above equation, adding $\log(ab)$ to both sides of the equation gives

$$\log A = -\frac{kt}{2.303} + \log A_0$$

which is Equation (7).

ANSWERS TO EXERCISES *Hydrolysis of a Tertiary Halide*

1.
$$[H^+] = \text{antilog}(-pH)$$
$$-pH = -2.47 = 0.53 - 3.00$$
$$\text{antilog}(0.53 - 3.00) = 3.4 \times 10^{-3} \ M$$

2. $\log 2.29 = 0.360$

3. See answer to Exercise 3 of Experiment 26.

4. $k = \text{slope} \times 2.303 = (1.67 \times 10^{-2}) \times 2.303 = 3.85 \times 10^{-2} \ \text{min}^{-1}$

$$k = \frac{3.85 \times 10^{-2} \ \text{min}^{-1}}{60 \ \text{sec/min}} = 6.42 \times 10^{-4} \ \text{sec}^{-1}$$

5. For a first-order reaction,

$$t_{1/2} = \frac{\ln 2}{k} = \frac{0.693}{3.85 \times 10^{-2} \ \text{min}^{-1}} = 18.0 \ \text{min}$$

Appendix II
Computer Programs

```
            BASIC PLUS PROGRAM FOR EXPERIMENT 1
            MEASURING VOLUME BY COUNTING DROPS


INPUT DATA ARE NUMBER OF TEN DROP INCREMENTS, TEMPERATURE OF
WATER IN DEGREES CELCIUS, DENSITY OF WATER IN G/ML, AND WEIGHTS
OF FLASK WITH EACH INCREMENT OF WATER.

OUTPUT DATA ARE TABLE OF VOLUME, DEVIATION FROM AVERAGE AND
SQUARE OF DEVIATION FROM AVERAGE FOR EACH TEN DROP INCREMENT;
AVERAGE VOLUME, AVERAGE DEVIATION, AND STANDARD DEVIATION.



 1 REM EXPERIMENT 1 - DATA TREATMENT
 2 REM MEASURING VOLUME BY COUNTING DROPS
 3 REM D = DENSITY OF WATER; V = VOLUME IN ML
 4 REM D1 = DEVIATION FROM AVERAGE
10 PRINT "EXPERIMENT 1 - MEASURING VOLUME BY COUNTING DROPS"
11 PRINT
12 PRINT "TYPE IN REQUESTED DATA WHEN TELETYPE PAUSES AFTER ?."
13 PRINT "    THEN PRESS RETURN.":PRINT
14 INPUT "HOW MANY 10 DROP INCREMENTS ";N
15 INPUT "TEMPERATURE IN DEG C = ";T
16 PRINT "DENSITY OF WATER AT ";
17 PRINT USING "##.#",T;
18 INPUT " = ";D
19 DIM D1(20), G(20)
20 PRINT
21 PRINT "NO. DROPS", "  WEIGHT OF"
22 PRINT "IN FLASK", "FLASK+H2O IN G"
23 FOR I=0 TO N
24 PRINT USING "###.",I*10;
25 PRINT TAB(14);
26 INPUT G(I)
27 NEXT I
28 PRINT:PRINT
30 FOR I=1 TO N
```

```
31 LET V1=V1+(G(I)-G(I-1))/D
32 NEXT I
33 LET V2=V1/N
40 FOR I=1 TO N
41 LET D1(I)=ABS((G(I)-G(I-1))/D-V2)
42 LET D2=D2+D1(I)
43 LET D3=D3+D1(I)^2
44 NEXT I
45 LET D4=D2/N:D5=SQR(D3/(N-1))
50 PRINT "NO.DROPS","VOLUME OF","DEVIATION","SQUARE OF"
51 PRINT "IN FLASK","INCREMENT","FROM AVG","DEVIATION"
52 PRINT
53 FOR I=1 TO N
54 PRINT USING "###",I*10;
55 PRINT TAB(15);
56 PRINT USING "#.####",(G(I)-G(I-1))/D;
57 PRINT TAB(29);
58 PRINT USING "#.####",D1(I);
59 PRINT TAB(43);
60 PRINT USING "#.#####",D1(I)^2
61 NEXT I
65 PRINT
66 PRINT "AVERAGE VOLUME = ";
67 PRINT USING "#.####",V2;
68 PRINT " MILLILITERS"
69 PRINT "AVERAGE DEVIATION = ";
70 PRINT USING "#.####",D4
71 PRINT "STANDARD DEVIATION = ";
72 PRINT USING "#.####",D5
80 END
```

BASIC PLUS PROGRAM FOR EXPERIMENT 2 PART II
CALIBRATION OF VOLUMETRIC GLASSWARE - BURET

INPUT DATA ARE NUMBER OF BURET READINGS (INCLUDING 00.00),
WATER TEMPERATURE IN DEGREES CELCIUS, DENSITY OF WATER IN
G/ML, BURET READINGS AND CORRESPONDING WEIGHTS OF FLASK +
WATER.

DATA TREATMENT GIVES TABLE OF CORRECTED WEIGHT, VOLUME AND
CORRECTION FOR EACH BURET READING.

```
1  REM EXPERIMENT 2 - DATA TREATMENT
2  REM CALIBRATION OF VOLUMETRIC GLASSWARE
3  REM PART II, BURET, ONLY
10 PRINT "EXPERIMENT 2 - BURET CALIBRATION":PRINT
11 PRINT "TYPE IN REQUESTED DATA WHEN TELETYPE PAUSES AFTER ?."
12 PRINT "    THEN PRESS RETURN":PRINT:PRINT
13 INPUT "HOW MANY BURET READINGS (INCLUDING 0.00) ";N
14 INPUT "TEMPERATURE OF WATER IN DEG C ";T
15 PRINT "DENSITY OF WATER AT ";
16 PRINT USING "##.#",T;
17 INPUT " DEGREES C ";D
18 DIM B(15), G(15), W(15)
20 PRINT:PRINT "   BURET", "WEIGHT FLASK"
21 PRINT "READING (ML)", " + H2O (G)"
22 FOR I=1 TO N
23 INPUT B(I)
24 PRINT TAB(14);
25 INPUT G(I)
26 PRINT
27 NEXT I
30 PRINT "   BURET", "CORRECTED", "VOLUME", "CORRECTION"
31 PRINT "READING (ML)", "WEIGHT(G)", " (ML)", "   (ML)"
32 FOR I=1 TO N
33 LET W(I)=(G(I)-G(1))+0.001*(G(I)-G(1))
34 PRINT USING "##.##",B(I);
35 PRINT TAB (14);
36 PRINT USING "##.##",W(I);
37 PRINT TAB(28);
38 PRINT USING "##.##",W(I)/D;
39 PRINT TAB(45);
40 PRINT USING "#.##",(W(I)/D-B(I))
41 NEXT I
50 END
```

BASIC PLUS PROGRAM FOR EXPERIMENT 16
HEAT OF FUSION

INPUT DATA ARE MOLE FRACTION OF NAPHTHALENE AND TEMPERATURE IN DEGREES CELCIUS.

DATA TREATMENT GIVES DELTA H, DELTA S, DELTA G AT 25.0 DEG C, STANDARD DEVIATION OF SLOPE, STANDARD DEVIATION OF INTERCEPT, AND DEVIATION OF EACH POINT FROM THE CALCULATED VALUE IN TERMS OF STANDARD DEVIATION.

```
1 REM EXPERIMENT 16 - DATA TREATMENT
2 REM ENTHALPY OF FUSION
4 REM X = MOLE FRACTION OF NAPHTHALENE; T = CELCIUS TEMPERATURE
5 REM S = SLOPE; C = ORDINATE INTERCEPT
6 REM D1 = APPROX STD DEVIATION OF SLOPE; V = VARIANCE OF (LNX)
7 REM D2 = APPROX STD DEVIATION OF INTERCEPT
10 PRINT "EXPERIMENT 16 - ENTHALPY OF FUSION":PRINT
11 PRINT "TYPE IN REQUESTED DATA WHEN TELETYPE PAUSES AFTER ?."
12 PRINT "    THEN PRESS RETURN.":PRINT:PRINT
13 INPUT "HOW MANY DATA POINTS ";N
14 DIM X(10), T(10), Z(10)
20 PRINT "POINT NO.", "MOL FRACTION", "DEG C"
21 FOR I=1 TO N
22 PRINT I;TAB(15);
23 INPUT X(I)
24 PRINT TAB(27);
25 INPUT T(I)
26 PRINT
27 NEXT I
30 FOR I=1 TO N
31 LET S1=S1+1/X(I):S2=S2+1/((T(I)+273.15)*X(I))
32 LET S3=S3+LOG(X(I))/X(I):S4=S4+1/((T(I)+273.15)^2*X(I))
33 LET S5=S5+LOG(X(I))/((T(I)+273.15)*X(I))
34 LET S6=S6+LOG(X(I))^2/X(I)
35 NEXT I
40 LET E=S1*S4-S2*S2:C=(S4*S3-S2*S5)/E:S=(S5*S1-S2*S3)/E
41 FOR I=1 TO N
42 LET V=V+((LOG(X(I))-C-(S/(T(I)+273.15)))^2)/(N-2)
43 NEXT I
44 LET D1=SQR(V*S1/E):D2=SQR(V*S4/E)
45 FOR I=1 TO N
46 LET Z(I)=ABS(S/(T(I)+273.15)+C-LOG(X(I)))/SQR(V)
47 NEXT I
48 PRINT:PRINT
50 PRINT "POINT NO.", "X", "LOG(10)X", "1/T"
51 FOR I=1 TO N
52 PRINT I;TAB(13);
53 PRINT USING "#.###",X(I);
54 PRINT TAB(29);
55 PRINT USING "#.###",LOG10(X(I));
56 PRINT TAB(39);
57 PRINT USING "#.#####",1/(T(I)+273.15)
58 NEXT I
60 PRINT:PRINT:PRINT
61 LET H1=-S*1.987E-3:H2=-S*8.314E-3
62 LET E1=C*1.987:E2=C*8.314
63 LET G1=H1-298.15*E1/1000:G2=H2-298.15*E2/1000
```

```
70 PRINT "DELTA H = ";
71 PRINT USING "#.##",H1;
72 PRINT " KCAL PER MOLE"
73 PRINT "      OR ";
74 PRINT USING "##.#",H2;
75 PRINT " KILOJOULE PER MOLE"
76 PRINT "DELTA S = ";
77 PRINT USING "##.#",E1;
78 PRINT " CAL PER MOLE PER DEGREE"
79 PRINT "      OR ";
80 PRINT USING "##.#",E2;
81 PRINT " JOULE PER MOLE PER DEGREE":PRINT
82 PRINT "DELTA G AT 25.0 DEG C = ";
83 PRINT USING "#.###",G1;
84 PRINT " KCAL PER MOLE"
85 PRINT "      OR ";
86 PRINT USING "#.##",G2;
87 PRINT " KILOJOULE PER MOLE":PRINT:PRINT
90 PRINT "STD DEVIATION OF SLOPE = ";
91 PRINT USING "##.#",D1
92 PRINT "STD DEVIATION OF INTERCEPT = ";
93 PRINT USING "#.##",D2
94 PRINT:PRINT "POINT NO.", "DEVIATION FROM CALC VALUE"
95 PRINT "               (TIMES STD DEVIATION)"
96 FOR I=1 TO N
97 PRINT I;TAB(20);
98 PRINT USING "##.##",Z(I)
99 NEXT I
100 PRINT "YOU MAY WANT TO CONSIDER NOT USING ANY POINT"
101 PRINT "  WITH A DEVIATION GREATER THAN 1.5 TIMES STD DEVIATION."
110 PRINT:PRINT
111 PRINT "LITERATURE VALUE OF DELTA H IS 4.558 KCAL PER MOLE"
112 PRINT "  OR 19.07 KILOJOULE PER MOLE."
120 END
```

BASIC PLUS PROGRAM FOR EXPERIMENT 26
DISSOCIATION OF AN IRON(II) COMPLEX

INPUT DATA ARE NUMBER OF DATA POINTS, TEMPERATURE IN DEGREES
CELCIUS, PERCENT TRANSMITTANCE, AND TIME IN MINUTES.

DATA TREATMENT GIVES RATE CONSTANT IN RECIPROCAL MINUTES FROM
INPUT DATA, LITERATURE VALUE AT THE SPECIFIED TEMPERATURE AND
DEVIATION OF EACH POINT FROM CALCULATED VALUE OF THAT POINT IN
TERMS OF STANDARD DEVIATION.

```
1 REM EXPERIMENT 26 - DATA TREATMENT
2 REM KINETICS OF DISSOCIATION OF AN IRON(II) COMPLEX
3 REM LEAST SQUARES FIT TO LN(A) = S*T + C
4 REM A = ABSORBANCE; T = TIME IN MINUTES; P = % TRANSMITTANCE
5 REM S = SLOPE; T2 = TEMPERATURE IN DEG C
6 REM V = VARIANCE OF (LN A); D = APPROX STD DEVIATION OF SLOPE
8 PRINT "EXPERIMENT 26 - IRON(II) KINETICS":PRINT
9 PRINT "TYPE IN REQUESTED DATA WHEN TELETYPE PAUSES AFTER ?."
10 PRINT "   THEN PRESS RETURN.":PRINT:PRINT
11 INPUT "HOW MANY DATA POINTS ";N
12 INPUT "TEMP IN DEGREES C ";T2:PRINT
13 PRINT "POINT NO.    %T       TIME(MIN)":PRINT
14 DIM P(16), T(16), A(16), Z(16)
15 FOR I=1 TO N
16 PRINT I;TAB(9);
17 INPUT P(I)
18 PRINT TAB(20);
19 INPUT T(I)
20 PRINT
21 NEXT I
22 FOR I=1 TO N
23 LET A(I)=2.000-LOG10(P(I)):S1=S1+1/A(I):S2=S2+T(I)/A(I)
24 LET S3=S3+LOG(A(I))/A(I):S4=S4+T(I)^2/A(I)
25 LET S5=S5+T(I)*LOG(A(I))/A(I):S6=S6+LOG(A(I))^2/A(I)
26 NEXT I
27 LET E=S1*S4-S2*S2:C=(S4*S3-S2*S5)/E:S=(S5*S1-S2*S3)/E
28 FOR I=1 TO N
29 LET V=V+(LOG(A(I))-C-S*T(I))^2/(N-2)
30 NEXT I
31 LET D=SQR(V*S1/E):T3=T2+273.15:K=1.192E21*EXP(-1.611E4/T3)
33 FOR I=1 TO N
34 LET Z(I)=ABS(S*T(I)+C-LOG(A(I)))/SQR(V)
36 NEXT I
37 PRINT
38 PRINT "POINT NO.","% T","ABSORBANCE","LOG(10)ABSORBANCE"
39 FOR I=1 TO N
40 PRINT I;TAB(14);
41 PRINT USING "##.#",P(I);
42 PRINT TAB(31);
43 PRINT USING "#.###",A(I);
44 PRINT TAB(47);
45 PRINT USING "#.###",LOG10(A(I))
46 NEXT I
47 PRINT:PRINT:PRINT
48 LET R=-S:PRINT "RATE CONSTANT = ";
49 PRINT USING "#.#####",R;
50 PRINT "  RECIPROCAL MINUTES":PRINT
```

```
51 PRINT "STD DEVIATION OF SLOPE = ";
52 PRINT USING "#.#####",D:PRINT
60 PRINT "POINT NO.","DEVIATION FROM CALC. VALUE"
61 PRINT "             (TIMES STD DEVIATION)"
62 FOR I=1 TO N
63 PRINT I;TAB(20);
64 PRINT USING "##.##",Z(I)
65 NEXT I
66 PRINT "YOU MAY WANT TO CONSIDER NOT USING ANY POINT OR POINTS"
67 PRINT " WITH A DEVIATION GREATER THAN 1.5 TIMES STD DEVIATION."
68 PRINT:PRINT:PRINT
69 PRINT "LITERATURE VALUE OF K AT ";
70 PRINT USING "##.#",T2;
71 PRINT "DEG C = ";
72 PRINT USING "#.#####",K
90 END
```

BASIC PLUS PROGRAM FOR EXPERIMENT 27
HYDROLYSIS OF A TERTIARY HALIDE

INPUT DATA ARE NUMBER OF POINTS, PH AT TIME INFINITY, TIME IN MINUTES AND CORRESPONDING PH.

DATA TREATMENT GIVES RATE CONSTANT IN RECIPROCAL MINUTES AND RECIPROCAL SECONDS AND THE STANDARD DEVIATION OF THE SLOPE.

```
1 REM EXPERIMENT 27 - DATA TREATMENT
2 REM HYDROLYSIS OF A TERTIARY HALIDE
3 REM T = TIME IN MINUTES; PH VALUES INDICATED BY P
4 REM HYDROGEN ION CONCENTRATIONS INDICATED BY H
5 REM S = SLOPE; D = APPROX STD DEVIATION OF SLOPE
10 PRINT "EXPERIMENT 27 - HYDROLYSIS OF A TERTIARY HALIDE":PRINT
11 PRINT "TYPE IN REQUESTED DATA WHEN TELETYPE PAUSES AFTER ?."
12 PRINT "    THEN PRESS RETURN":PRINT:PRINT
13 INPUT "HOW MANY DATA POINTS ";N
14 INPUT "PH AT TIME INFINITY ";P9:PRINT
15 DIM P(25), H(25), P1(25), T(25), Z(25)
16 PRINT "POINT NO.", "TIME(MIN)", "PH":PRINT
17 FOR I=1 TO N
18 PRINT I;TAB(15);
19 INPUT T(I)
20 PRINT TAB(26);
21 INPUT P(I)
22 PRINT
23 NEXT I
30 LET H1=EXP(-P9*2.3026)
31 FOR I=1 TO N
32 LET H(I)=EXP(-P(I)*2.3026)
33 LET P1(I)=(H1-H(1))/(H1-H(I))
34 LET S1=S1+1:S2=S2+(T(I)-T(1)):S3=S3+LOG(P1(I))
35 LET S4=S4+(T(I)-T(1))^2:S5=S5+(T(I)-T(1))*LOG(P1(I))
36 LET S6=S6+LOG(P1(I))^2
37 NEXT I
40 LET E=S1*S4-S2*S2:C=(S4*S3-S2*S5)/E:S=(S5*S1-S2*S3)/E
41 FOR I=1 TO N
42 LET V=V+(LOG(P1(I))-C-S*(T(I)-T(1)))^2/(N-2)
43 NEXT I
44 LET D=SQR(V*S1/E)
45 FOR I=1 TO N
46 LET Z(I)=ABS(S*(T(I)-T(1))+C-LOG(P1(I)))/SQR(V)
47 NEXT I
50 PRINT "POINT NO.", "[H]", "(  )/(  )", "LOG10 (  )/(  )"
51 PRINT
52 FOR I=1 TO N
53 PRINT I;TAB(12);
54 PRINT USING "#.###^^^^",H(I);
55 PRINT TAB(30);
56 PRINT USING "#.###",P1(I);
57 PRINT TAB(46);
58 PRINT USING "#.###",LOG10(P1(I))
59 NEXT I
60 PRINT:PRINT
65 PRINT "RATE CONSTANT = ";
66 PRINT USING "#.###^^^^",S;
67 PRINT " RECIPROCAL MINUTES"
```

```
70 PRINT "RATE CONSTANT = ";
71 PRINT USING "#.###^^^^",S/60;
72 PRINT " RECIPROCAL SECONDS":PRINT
73 PRINT "STD DEVIATION OF SLOPE = ";
74 PRINT USING "#.##^^^^",D
75 PRINT:PRINT
80 PRINT "POINT NO.", "DEVIATION FROM CALC. VALUE"
81 PRINT "                    (TIMES STD DEVIATION)"
82 FOR I=1 TO N
83 PRINT I;TAB(18);
84 PRINT USING "##.##",Z(I)
85 NEXT I
86 PRINT "YOU MAY WANT TO CONSIDER NOT USING ANY POINT OR POINTS"
87 PRINT "   WITH A DEVIATION GREATER THAN 1.5 TIMES STD DEVIATION."
90 END
```

Appendix III
Tables

CONSTANTS

Avogadro's number	6.023×10^{23} mol^{-1}
Faraday constant	9.6487×10^{4} coulomb/mol
Gas constant	8.206×10^{-2} l atm/mol K
	1.987 cal/mol K
	8.314 joule/mol K
Molar volume of gas at STP	22.41 l/mol
Ice point temperature	273.150 K
$\pi = 3.1416$	
$e = 2.7183$	

CONVERSION FACTORS
(To Convert)

From	To	Multiply By
kcal	joule	4.184×10^{3}
joule	kcal	2.390×10^{-4}
atm	torr	760
torr	atm	1.316×10^{-3}
angstrom unit	nm	0.1
inch	cm	2.540
cm	inch	3.937×10^{-1}
pound	g	4.536×10^{2}
g	pound	2.205×10^{-3}
$\log_{10} X$	$\log_{e} X$	2.303

SALT SOLUBILITIES IN WATER

Ion	Generally	Common Exceptions
NO_3^-	soluble	none
Cl^-	soluble	$AgCl$, Hg_2Cl_2, $PbCl_2$*
SO_4^{2-}	soluble	$CaSO_4$*, $BaSO_4$, $HgSO_4$, $PbSO_4$, Ag_2SO_4*, Hg_2SO_4
CO_3^{2-}	insoluble	Group IA and NH_4^+ compounds
PO_4^{3-}	insoluble	Group IA and NH_4^+ compounds
OH^-	insoluble	Group IA, $Ca(OH)_2$*, $Ba(OH)_2$
S^{2-}	insoluble	Group IA, IIA and NH_4^+ compounds
Na^+	soluble	none
NH_4^+	soluble	none
K^+	soluble	$KClO_4$*

*Borderline case, commonly called slightly soluble.

PERIODIC TABLE OF THE ELEMENTS

PERIODS	IA	IIA	IIIB	IVB	VB	VIB	VIIB	VIII			IB	IIB	IIIA	IVA	VA	VIA	VIIA	0
1	1.008 **H** 1																	4.003 **He** 2
2	6.941 **Li** 3	9.012 **Be** 4											10.81 **B** 5	12.011 **C** 6	14.007 **N** 7	15.999 **O** 8	18.998 **F** 9	20.179 **Ne** 10
3	22.990 **Na** 11	24.305 **Mg** 12					TRANSITION ELEMENTS						26.982 **Al** 13	28.086 **Si** 14	30.9738 **P** 15	32.06 **S** 16	35.453 **Cl** 17	39.948 **Ar** 18
4	39.102 **K** 19	40.08 **Ca** 20	44.956 **Sc** 21	47.90 **Ti** 22	50.941 **V** 23	51.996 **Cr** 24	54.938 **Mn** 25	55.847 **Fe** 26	58.933 **Co** 27	58.71 **Ni** 28	63.546 **Cu** 29	65.37 **Zn** 30	69.72 **Ga** 31	72.59 **Ge** 32	74.922 **As** 33	78.96 **Se** 34	79.904 **Br** 35	83.80 **Kr** 36
5	85.468 **Rb** 37	87.62 **Sr** 38	88.906 **Y** 39	91.22 **Zr** 40	92.9064 **Nb** 41	95.94 **Mo** 42	98.906 **Tc** 43	101.07 **Ru** 44	102.906 **Rh** 45	106.4 **Pd** 46	107.868 **Ag** 47	112.40 **Cd** 48	114.82 **In** 49	118.69 **Sn** 50	121.75 **Sb** 51	127.60 **Te** 52	126.904 **I** 53	131.30 **Xe** 54
6	132.906 **Cs** 55	137.34 **Ba** 56	**La-Lu** 57-71	178.49 **Hf** 72	180.948 **Ta** 73	183.85 **W** 74	186.2 **Re** 75	190.2 **Os** 76	192.22 **Ir** 77	195.09 **Pt** 78	196.967 **Au** 79	200.59 **Hg** 80	204.37 **Tl** 81	207.2 **Pb** 82	208.981 **Bi** 83	(209) **Po** 84	(210) **At** 85	(222) **Rn** 86
7	(223) **Fr** 87	226.025 **Ra** 88	**Ac-Lr** 89-103	(261) [**Rf**] 104	(260) [**Ha**] 105													

Lanthanum Series

138.906 **La** 57	140.12 **Ce** 58	140.908 **Pr** 59	144.24 **Nd** 60	(145) **Pm** 61	150.4 **Sm** 62	151.96 **Eu** 63	157.25 **Gd** 64	158.925 **Tb** 65	162.50 **Dy** 66	164.930 **Ho** 67	167.26 **Er** 68	168.934 **Tm** 69	173.04 **Yb** 70	174.97 **Lu** 71

Actinium Series

(227) **Ac** 89	232.038 **Th** 90	231.031 **Pa** 91	238.029 **U** 92	237.048 **Np** 93	(244) **Pu** 94	(243) **Am** 95	(247) **Cm** 96	(247) **Bk** 97	(251) **Cf** 98	(254) **Es** 99	(253) **Fm** 100	(256) **Md** 101	(253) **No** 102	(257) **Lr** 103

Numbers below the symbol of the element indicate the atomic numbers. Atomic masses, above the symbol of the element, are based on the assigned relative atomic mass of ^{12}C = exactly 12; ( ) indicates the mass number of the isotope with the longest half-life. [ ] indicates not officially approved.